T

CARE AND MANAGEMENT

OF

STATIONARY STEAM ENGINES

A Practical Handbook for Men-in-charge

BY

CHARLES HURST

AUTHOR OF "VALVES AND VALVE GEARING," "HINTS ON STEAM ENGINE DESIGN
AND CONSTRUCTION," ETC.

WITH THIRTY-ONE ILLUSTRATIONS

First published 1906
Reprinted 1993
© TEE Publishing ISBN 1 85761 028 8

PREFACE.

THIS little volume contains the substance of articles published in serial form in *Mining Engineering:* hence its special reference to engines used in and about mines, and collieries in particular. The articles were mostly written in the winter evenings of 1901-2, straight from the writer's daily experience, without reference to any books whatever. The work makes no pretensions to profundity, but aims at being *practical*. No special precautions have been taken to prevent its being interesting.

CHARLES HURST.

TRENT VIEW,
BURTON JOYCE, NOTTS.
October, 1902.

THE CARE AND MANAGEMENT

OF STATIONARY STEAM ENGINES

THE

CARE AND MANAGEMENT

OF

STATIONARY STEAM ENGINES

CHARLES HURST

First Published 1906
Reprinted 1992
© TEE Publishing ISBN 1 85761 028 8

CONTENTS.

———•———

CHAPTER VI.

THE
CARE AND MANAGEMENT
<small>OF</small>
STATIONARY STEAM ENGINES.

CHAPTER I.

WATER IN CYLINDERS AND LEAKAGES.

INNUMERABLE books and papers have been written on the design and construction of steam engines, but so far as the writer is aware the literature dealing with the care and management of steam engines is distinctly scanty. It is said that a wise head makes a still tongue, so perhaps as a rule the man who has the skill to take charge of engines successfully is too wise to scatter his wisdom broadcast. Now although the writer has never been in charge of any particular engine for an extended period, a considerable part of his life has been spent in designing engines that would escape the censure of the men who have them under their care, in repairing breakdowns, and in making alterations to defective arrangements, so that he has some knowledge of the views of engine minders and their methods, and believes that the present work may prove useful to some of those who, without a long experience, are held responsible for the working of engines, as well as to those who may have to design them.

The greatest enemy the attendant engineer has to deal with is the presence of water in the cylinders; the forms of the latter in many cases appear to be designed without regard to this important matter. This much may be said at the outset, that the less water lodging anywhere in the cylinder and valve chests the better, and mainly for this reason the Corliss cylinder with the exhaust valves at the bottom corners of the cylinder is superior in efficiency to the slide valve cylinder. On first taking charge of an engine the attendant, after a general inspection, should consider carefully the form of the cylinder and valve chest, and note where it is possible for water to lodge. He should then ask himself what the effects of water lodging at any particular point will be. For instance, take the case of a slide valve cylinder which is not fitted with escape valves, a common case in small engines of old date. In such a case a small quantity of water, although by no means conducive to economy, would not be specially dangerous, because there would be room between the piston when at the end of the stroke and the cylinder cover for the water to lodge, but should the quantity of water be too great, disaster will surely follow, for the piston will come upon the water as upon a solid mass, and either the cover, piston, piston rod, or connecting rod will break, or the crank pin will be bent. It may be argued that the steam port is always open to the cylinder and the water can find an outlet by forcing the valve off its seat, but when the piston strikes the water it finds the latter inert, and before the velocity of the piston can be communicated to the water the mischief is done. The attendant should therefore pay great attention to the cylinder drain cocks, especially at starting; and during running he should open them every one or two hours as he may find advisable. When escape valves are provided they should be lifted every day to ascertain that they are not sticking or binding on their face.

As the engineer surveys the engine in detail, he may find many places where water will lodge, such as in the pipes between the high and low pressure cylinders of compound engines, in the stop valve, at the bottom of the exhaust pipes, and possibly in the valve boxes. Such places should be noted, and as time and opportunity offer, holes should be tapped and drain cocks fitted in such positions as may be suitable. A few days' running will show how much water accumulates at the various points, and from this an opinion may be formed as to the frequency with which the several cocks should be opened.

The presence of water in the main steam pipes is well known to be dangerous at starting. When the junction valve on the boiler is opened a rush of steam travels along the pipes, and, carrying along any water that may be in its course, projects it as a solid mass against the first obstacle. This action is commonly known as water-hammer, and has been the cause of many fatal disasters. The engineer should therefore be very careful to open the junction valve very gradually, especially where the pipes to the engine have a long and circuitous route, with frequent rises and falls where water may lodge. Similar precautions are necessary on opening blow-through valves, warning valves, and in fact where any steam valve opens into a long range of pipes. A steam trap is better than an ordinary drain cock, but the expense is often prohibitive. The action of a trap being entirely automatic, the engineer is relieved of considerable anxiety. The principle of the action of most traps is similar to that of a valve on an automatic filling tank. A ball float is fitted inside the trap, and as the water accumulates the float rises and opens a valve for the escape of water. Others, again, depend on the expansion of metals for their action. When full of water they are constructed so that a valve at the end of a long copper or gun-metal rod or tube is slightly off its face, thus allowing the water to flow away, but as the water gives place to steam, the rod expands by reason of the higher temperature of the steam, and so closes the valve. The MacCracken trap is made on this principle, and has proved very satisfactory and reliable.

Whatever kind of traps are fitted they require periodical examination to insure that the valves are not sticking or that the passages are not choked up. If convenient, all the traps from the engine should be taken by separate pipes to a tank fixed in some convenient position, so that the total water made by the engine can be measured, and if found excessive steps can be taken to ascertain the cause.

Leakages and blowing joints are a continual source of annoyance and anxiety to engineers. In a range of steam pipes want of elasticity is the principal cause. The joints may be made of good material and properly screwed up, but when the steam is passed through the range the expansion may tend to open some of the joints and make them impossible to keep tight. The remedy is to put in an expansion joint, or else re-arrange the pipes so that the effect of expansion is to twist one flange of the pipe on the face of the other, instead of opening the joint. The following diagrams which show in outline a range of pipes from a boiler to the cylinder of an engine will explain matters clearly. In the first arrangement (Fig. 1) trouble would in all probability be experienced by some of the joints blowing. In the next (Fig. 2), however, the effect of the expansion is to twist the flanges B on the flanges C rather than open the joints D. The whole range in fact is more elastic. Should space not permit of an arrangement of this nature being adopted, an expansion joint might be inserted (Fig. 3). This consists of a copper ring of **U** section stiffened at the flange by wrought-iron rings.

Leaky glands are a very common complaint, and with steam of high pressures they are very wasteful; hence metallic packings are an advantage. A soft packing is generally satisfactory for a time, but as the rods wear down

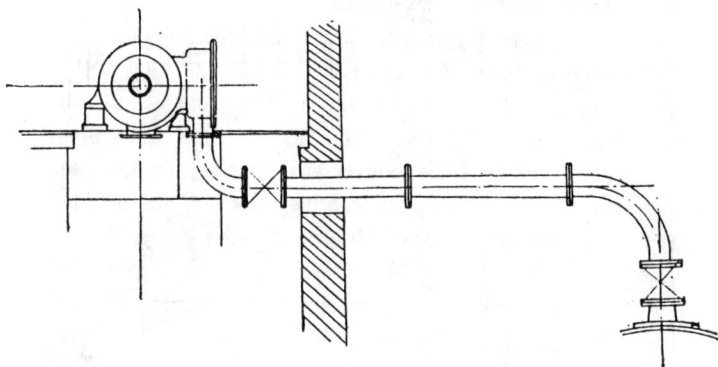

Fig. 1.

on the bushes the glands require screwing up very tightly to prevent escape of steam, and considerable friction is the result. In addition to this there is the frequent renewal of the packing to be taken into account, a somewhat expensive item. It is therefore a good investment to place metallic packings on all the spindles and rods of an engine, although the first cost is somewhat excessive, the price for a good metallic packing being somewhere about seventy shillings per inch of the diameter of the rod.

An effective arrangement for preventing leakage at the

glands is given on page 8 (Fig. 4), which consists of a ring of brass in which are formed suitable channels and passages. This ring is inserted midway in the stuffing box, as shown, and a water supply arranged to fill the spaces in the ring. This device is very effectual in preventing leakage, though it is not suitable for anything but a pump or compressor.

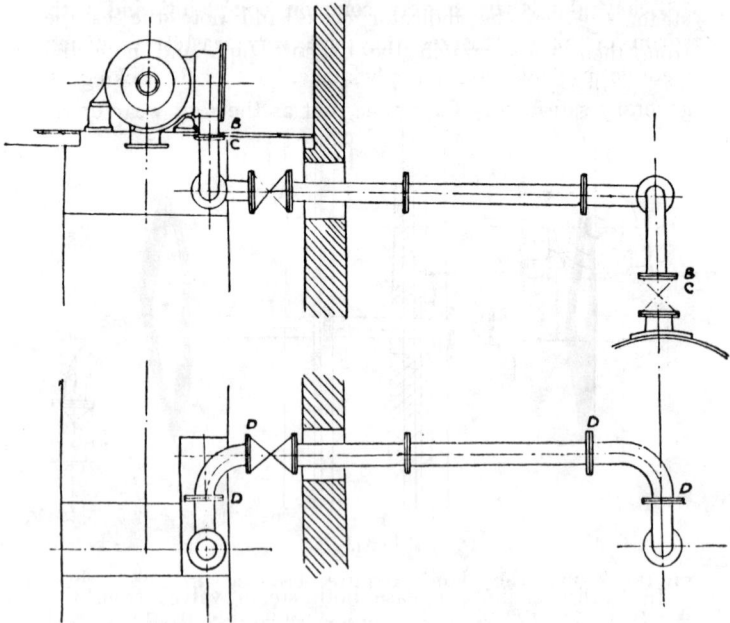

Fig. 2.

It is of special advantage in horizontal air pump rods, where a slight leakage is detrimental to the efficiency of the vacuum.   A lantern brass can be fitted to any ordinary stuffing box, and where trouble through leakage has occurred it is recommended as a good remedy.

Leaky valves are often very wasteful, because the leakage is not apparent, and may go on for years without being discovered. There are several ways of testing for a leaky valve. Sometimes the leakage is so great that it may be noticed in the indicator diagram. In the case of slide valve engines the best way to test for leakage is to bar the engine round until the valve has closed over both steam ports of the cylinder. Then scotch the engine and turn on the steam. Open the indicator cocks and note the escape from them. Also watch the exhaust pipe and note the escape there.

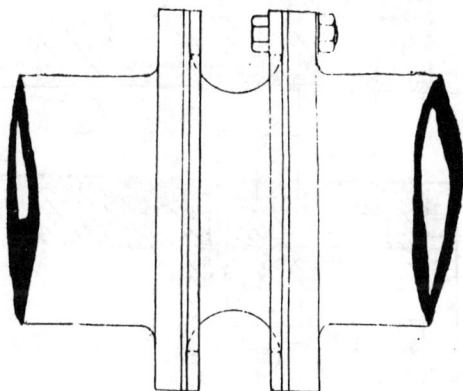

Fig. 3.

In Corliss engines, release both steam valves from the catches, when they will be closed over the ports, and open the indicator cocks, and any considerable escape past either steam valve will be shown. To test the exhaust valves, bar round until one valve is fully closed, scotch the engine, and open the steam valve on the same end of the cylinder as the exhaust valve being tested, having previously released the catches on the other side. Then turn on steam at the stop valve, gradually, and watch the exhaust pipe for leakage. Turn off the steam, bar round till the other exhaust valve is closed, and repeat the operation at that end of the cylinder.

In compound engines it is not so easy to test in the high pressure cylinder because the exhaust pipes form the steam pipes of the low pressure cylinder, and are therefore closed. The best means of testing for a leak in this case is to insert a small trumpet pipe in the exhaust, close to the valves, and provided with a stop cock. Turn the trumpet mouth

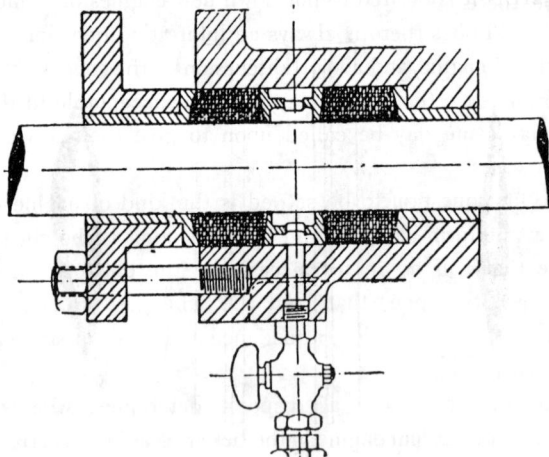

Fig. 4.

in the direction to catch the leakage, and open the tap. To test the low pressure valves of condensing engines, proceed in a similar manner, using a trumpet pipe, and arrange a small pipe to lead from the boiler steam pipes to the low pressure cylinder so as to obtain a pressure in the cylinder well above the atmosphere. If blow-through pipes or warning valves are provided, however, it will be unnecessary to provide the pipe leading from the boiler steam pipes to the cylinder.

In testing for leaky pistons the best way is to scotch the engine, take off one cover and admit steam gradually to the other side of the piston. When the cover is not removed it is difficult to say how much of the escape from the indicator cocks is due to a leaky piston and how much to the valve, but when the cover is removed the escape can be actually seen. Note also whether the steam escapes from the top of the cylinder, as is usual, or equally all round the edge. If escape should be only at the top it denotes that the packing is in fair condition, but the cylinder or piston has worn oval, and the latter dropped. Should the escape be all round, the packing is defective.

Sometimes the piston and valves may be tight at one part but leaky at others; so that they should be tested for tightness at various points in their respective movements. Great care should be taken to scotch the engine securely, and before turning on the steam at the stop valve it must be carefully considered what position the valve occupies, and whether the effect of turning on the steam will produce dangerous results. A little consideration will show that in order to test the piston throughout the whole stroke it may be necessary to uncouple the eccentric and secure the valve in one position, otherwise the steam port on that side of the cylinder under observation would be uncovered, with results that might be dangerous, and would certainly prevent accurate observations being taken.

The main stop valve, if of the ordinary screw-down mushroom type, is very liable to leakage, due sometimes to grit getting upon the seat, and sometimes to the pressure distorting the valve. Another frequent cause arises from the unequal expansion of the valve and seat, especially in double-beat or crown valves of the type frequently fitted to winding engines. In order to test for a leak, first close the junction valve on the boiler, then bar the engine round until the steam port at one end of the cylinder is wide open, screw down the stop valve and open the junction valve gradually, so as to avoid water-hammer. Open indicator cocks and watch for escape; or for more accurate observation remove the cylinder cover and watch the steam port. Take great care that whilst the cover is off no one has access to the stop valve handle.

Leakages into the condenser and air pump are not easily detected, because there is no outward and visible sign; yet the extent to which a small leak will reduce the degree of vacuum is surprising. Leakage commonly occurs at the stuffing box of the cylinder. Sometimes the joints of the flanges in the pipes from the cylinder to the condenser are not properly made, and sometimes leakage may occur through a spongy or defective casting of the condenser, or the pipes in connection therewith.

All suspicious places should be tested by holding the flame of a candle or taper, and with a little experience and care the slightest leak will be detected. Apply this test to the vacuum gauge connections, injection valve, and the joint of the foot valve cover. Lubricator fittings, indicator taps, escape valves, and similar fittings on the low pressure cylinders should also be tried if the engine is condensing and the steam pressure in the low pressure cylinder does not reach much above atmospheric pressure.

CHAPTER II.

KNOCKS AND THEIR CAUSES.

NOTHING is more annoying to an engineer who has
any pride in his engines than a knock in the journals,
or the cylinder, and it is proposed in this chapter to treat of a
few of the causes and their remedies.

Generally, the cause of a knock in the bearings of an
engine is the want of proper adjustment in the bearings,
but in high speed engines it may arise from want of correct
setting of the valve gear, or bad design in the moving
parts. As usually constructed, it is a tedious matter to adjust
the brasses of a large engine. First the cap has to be taken
off and the brasses or liners filed, and then tried on the
bearing; and this process may have to be repeated several
times before satisfactory adjustment is obtained. If, however, the brasses of connecting rods, valve rods, etc., were
made as shown in the annexed figure (Fig. 5), adjustment
would be a much simpler matter. Referring to the illustration, the movable part of the brass will be seen to extend
for less than half the circumference of the bearing, and as
the pin or the brasses wear, the adjustable step is able to
follow up the wear without any filing or scraping whatever.
The above arrangement has proved very satisfactory in
practice, and whenever new brasses are required it is recommended if the case permits of its being employed.

The crank pin is the journal that most frequently knocks,
the wear upon it being severe and subject to continual

reversals. After the crank pin, the crosshead brasses will
probably require the most frequent attention, and then the
bearings of the valve gear. In tightening up the various
journals great care is needed to avoid making them too tight,
a state of things that may prove more serious than a knock,
though not so conspicuous. No definite rules can be laid
down as to the amount of tightness necessary for good working, that being a question to be settled by the intelligence
of the engineer in charge and the particular circumstances
which obtain; but a common plan is to adjust the cotter

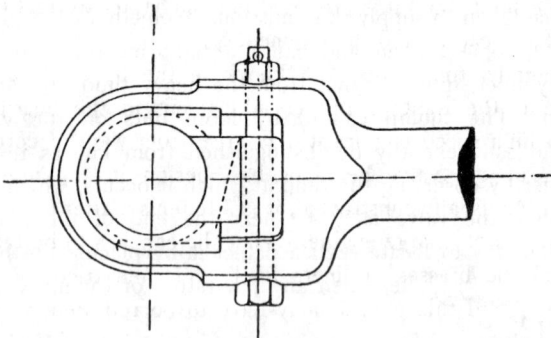

Fig. 5.

or wedge nut by means of the fingers only, and then lock
in position. This insures that no great pressure is put upon
the journals. After such adjustment the engine may be
started, and should there still be a knock a spanner may be
used with care; and if after this adjustment the brass should
then begin to heat, there is probably some other cause than
too tight adjustment, possibly grit in the bearings, want of
true alignment or proper lubrication. When replacing the
brasses they should be carefully wiped free from all filings,
grit, and dust, and the pin also cleaned in a similar manner.

So far, the knocks that have been mentioned are on the

external parts of the engine, and are easily located, but knocks in the cylinder and valve chests are sometimes difficult to account for. Sometimes it is the piston nut that works loose, a most dangerous state of things, and generally accompanied by a dull heavy thud towards the end of the stroke.

The engineer should at once satisfy himself on this matter, or he may have the cylinder covers knocked out, and if on examination the piston is found to be loose, and there are no means of locking or securing the nut, steps should be at once taken to supply the omission. Sometimes the piston rings become broken and cause a light metallic rattle in the cylinder, quite distinct from the heavy thud of a loose piston. The thump caused by loose nuts on the valve spindle can generally be distinguished from knocks in the cylinder by observing the time at which it occurs, which will be when the valve is at or near the end of the stroke. Another way to locate such a defect is by placing the hand on the valve spindle, when any jar, rattle, or thump will be readily felt.

When a valve lifts from its face, due perhaps to water in the cylinder, excessive compression, and often from reasons which can scarcely be traced, the result is a sharp hammer, often occurring towards the end of the stroke. Various remedies will suggest themselves, such as putting a spring behind the valve, or cutting a strip off the inside or exhaust edge of the valve, thereby reducing compression.

The knocks above mentioned are those arising for the most part from wear or neglect, but in addition there may be some that have a more subtle origin, the remedies for which are not so apparent. One of the chief causes of knocks not due to wear or the looseness of any of the parts is the inertia of the reciprocating parts, such as the piston rod, crosshead, piston and connecting rod. When steam is admitted to the cylinder a portion of the work of the steam is expended in overcoming the inertia of the moving parts. These reciprocating parts then acquire a velocity which carries them along, and towards the end of the stroke they may be checked by the crank pin, thus giving up the work they absorbed at the beginning of the stroke. Now it will be understood that unless there is a cushion or resistance offered to the piston towards the end of the stroke the reciprocating parts will be thrown with considerable force upon the crank pin and a shock be felt.

The worst case that could occur is that of a cylinder taking steam the whole length of the stroke, having a high piston speed, and no compression. It would be almost an impossibility for such an engine to run quietly, because the inertia of the reciprocating parts would be thrown full on the crank pin without check. If, however, instead of the above conditions the steam were cut off at say one-third of the stroke, and compression on the other side of the piston commenced at about 80 per cent. in the stroke a much quieter running would result, because not only is the pressure of the steam gradually decreasing on one side of the piston, but a cushion of steam formed on the other side. In addition to the quieter running there would also be an economy arising from the compression. In engine design it is always the aim of the designer to have compression at least equal to overcome the inertia of the moving parts, so that they are brought gently to rest at the end of the stroke, whereby smooth running is obtained.

From what has been said it will be inferred that some times engines can be made to run much easier by increasing the amount of compression, and when a noisy engine is met with the engineer should consider the probability of the knock being due to a want of it. In a slide-valve engine more compression can only be got by screwing lap strips on the exhaust edges of the valve, but in Corliss engines it may be altered by shifting the eccentric and slightly altering the valve rods.

The question of inertia cannot be fully understood unless the reader has some knowledge of indicator diagrams, and as it is proposed to treat of these in a subsequent chapter, it will be necessary to revert to the present subject later.

CHAPTER III.

LUBRICATION.

IN considering the lubrication of steam engines it will be convenient to deal first with internal lubrication, that is of such parts as the piston and valves. The practice of employing a sight-feed lubricator is now almost universal, except for small engines running intermittently. The chief point in the working of the sight-feed lubricator is to know what is the most economical rate of feed. Most lubricators of this type can be adjusted to give from one to 140 drops per minute, so that there is room for a wide error. It would be absurd to give any rule for the rate of feed, that depending on many circumstances, but the following is suggested as a means of reducing the oil consumption to a minimum :— Regulate the feed so that the supply is greater than circumstances would seem to demand, and each day reduce the rate of feed by one drop per minute : note carefully the running of the engine, paying particular attention to the noise made by the valves and pistons. In all probability there will be a point at which the engine sensibly depreciates in its running, which denotes that the oil supply is insufficient. When that point has been reached *increase* the rate of feed by, say, three or four drops per minute, and settle upon this as the rate of feed. Of course, should the load of the engine vary considerably the feed must be adjusted to the best advantage, a heavily loaded engine requiring a proportionally increased feed. In the lubricators for the external parts a quicker feed than necessary is little or no loss if proper measures be taken to collect and sieve the oil; but with internal lubrication whatever passes into the valve boxes or cylinders cannot be used again, hence it becomes desirable to know the minimum rate of feed that will give efficient lubrication. In charging the lubricators great care should be taken that the oil is free from all foreign substances, or the pipes and connections will become choked up.

In addition to the sight-feed lubricator it is usual to provide oil cups so that oil can be poured directly on the valve faces and cylinder barrel. This provision is useful on starting engines after they have stood for a time. A slight rust generally forms on the cylinder walls and valve faces, and when the engine is started the valves and pistons are liable to groan and move stiffly for a time, unless a flush of oil can be directed upon the working surfaces. Care should be taken that the taps and connections of all cylinder and valve chest oil cups are tight, or steam will enter the cup and if any oil has been left in from the previous charge it will become frothy. This is no detriment to the oil itself, but it fills the cup and prevents the lubricator being filled properly.

In the lubrication of the exterior parts of engines there are three distinct systems, each having its own special advantages: the ordinary drop system in which the oil is poured into cups or syphons and allowed to drop upon the bearings either by the regulation of a tap or by the action of a wick; the splash system in which the connecting rod end and crank and crank pin dash into a well of oil in the crank pit, and thus throw the lubricant over the working parts of the engine;

c

and forced lubrication in which the oil is forced upon the bearings by means of a small pump generally worked from the main crank shaft by an eccentric. The drop system is usually applied to large, slow-running engines, such as air-compressors, hauling, winding, pumping and fan engines. The splash and forced systems are chiefly confined to high speed, continuous-running engines such as are made for electric driving. The first system is old, the others comparatively recent. The drop system is applicable to all kinds of open engines; the other systems require the moving parts to be enclosed.

In the drop system clean worsteds and wicks are essential to proper lubrication. Whenever any part of the engine is taken asunder the oil grooves should be carefully scraped and wiped clear of all dirt, and if on examination it appears that oil does not reach any part of the journal a groove may be cut in the brass to conduct the oil to the desired point.

A journal may heat from three causes : want of alignment, too tight adjustment of the brasses, or improper and insufficient lubrication. The second cause of heating will occur after adjustment, and is easily recognised, but some skill is required to discriminate between the other two causes of trouble. Should an inspection of the brasses show them scored at the bottom of the journal at one end, and at the top of the other end, it is certain that want of alignment has something to do with the trouble; on the other hand, scoring across the whole length of the bearing would point to improper lubrication as the cause.

It is important that oil should not get on the foundations, the effect being to destroy the nature of the joints in the masonry, so that with the vibration of the machinery the stones become loose and the foundations sink. This is a serious matter, and in some cases has broken the bedplates and ruined the brasses in the main journal. Drip trays should therefore be placed under all parts from which oil is likely to drip. If the main pedestals are not provided with oil catchers steps should be taken to supply the defect. A good galvanized sheet-iron dripper can be bought very cheaply, and for most jobs is quite satisfactory; though if appearance is to be studied a brass dripper will look better. Splashers may also be required round the crank pin. Planished steel with brass beading makes the best and neatest crank splasher. In winding engines the drum splashers are usually made of pine boarding, which is the best where a large surface has to be shielded.

For the main bearings of important engines which cannot be stopped except at great inconvenience, it is a good plan to arrange for a water supply to be directed on the journals. This provision will sometimes enable the engine to run for some time under trying circumstances. The fan engines at the Wigan Coal and Iron Company's Broomfield pit have a well-arranged water service for the main journals which will recommend itself as being efficient, cheap, and neat. The handrailing round the engine consists of wrought-iron pipes which are connected to the water main. At convenient points taps are fitted, and a stream of water can readily be directed where desired. The use of a water stream, however, is attended with danger unless care is exercised. The full stream should never be turned suddenly upon a hot neck, or it may start a surface crack, not very serious at first, perhaps, but liable to develop into a serious flaw. The proper method is to start the stream very gradually, increasing the flow by small increments.

In the lubrication of piston rods, valve spindles and the like, the glands are sometimes provided with recesses, which form oil wells; and the oil is syphoned on to the rod with a worsted in the ordinary way. When such provision is not made a good plan is to twist a piece of hemp or spun yarn round the rod, fastening it in position by the gland studs. A cup should then be arranged to drip on the hemp. This arrangement, which is shown by the figure, insures that the whole circumference of the rod is lubricated (see Fig. 6).

The great advantage of the splash system is that it requires little attention. As long as enough oil is kept in the crank

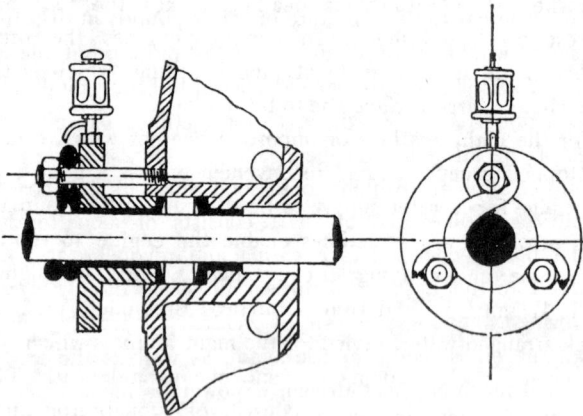

Fig. 6.

chamber to give a splash the lubrication will look after itself, provided the crank chamber is well cleaned of sand and grit, and the oil and grooves clean. Similar precautions apply to the forced system. The pressure gauge in connection with the oil pump will indicate whether the circulation is vigorous or otherwise. The failure of the system may occur through insufficient oil being in circulation, so that the suction of the pump draws air instead of oil. Sometimes one of the main bearings may not be closely fitted, in which case the whole supply of oil from the pump may escape at that one bearing, the others not receiving a supply. Another cause of failure may be sticking of the pump clacks due to foreign matter in the oil, such as bits of waste, filings, chippings, or sand from the enclosed parts of the engine.

Both the splash and forced systems are economical in oil consumption, because once charged no fresh oil is required for weeks. As ordinarily arranged, nothing could be more wasteful than the drop system. The oil is frequently allowed to drop on the bearing and afterwards flow "whither it listeth," whereas the expenditure of a few pounds in drip tins and sieves would prove about one of the most profitable investments it is possible to make. The oil bill for a moderate-sized engine frequently equals and sometimes exceeds the wages of the attendant engineer.

Another appliance which would soon repay its first cost is a centrifugal oil separator. It consists of a perforated circular basket enclosed in a casing and revolving at great speed. Into this basket are thrown all pieces of oily waste, from which the oil is drawn by centrifugal force. When sieved, the oil is ready for use again, as well as the waste. One small machine, hand-driven, if power be inconveniently situated, will suffice for the whole plant of a colliery, and would save its first cost in twelve months.

CHAPTER IV.

MISCELLANEOUS MATTERS.

HERE are several points of importance connected with the care of steam engines which do not come under any particular heading. In this chapter it is proposed to deal with some of these miscellaneous matters; and though the present article may appear to be of a scrappy and disjointed nature it is difficult to avoid such an effect where the subjects are so numerous and dissimilar.

When large engines are first set to work the engineer usually has a busy time. One important matter which is frequently neglected, is the fly-wheel or rope-pulley bolts. These should all be carefully tightened up, and for the first week or two they should be examined at the end of each day's run, and tried with a spanner to see if they have slacked back. Most engines are stiff at first, and until the working parts know each other, as engineers express it, they require liberal doses of oil. Those in superintendence of the erection will do well to see that all fitting parts are well greased or oiled as they are put together. This will save much annoyance, and possibly some strong language later on, when the engine is being overhauled. This remark applies especially to the nuts of the pistons, the crosshead cotters, the cylinder and pedestal feet bolts, connecting-rod bolts, and slide-bar bolts. It is very important that, before the cylinder and valve-chest covers are bolted on, careful inspection should be made with a view to discovering any loose sand in the ports and passages, which may have been left in through

careless dressing and fettling in the foundry. Any
sand left in will in time get on the valve faces and the
cylinder barrel and score them badly.

Hammers, chisels and blocks of wood have sometimes been
left in the cylinders, valve chests, and air pumps with
disastrous results. This is a matter which should always
be guarded against by thorough inspection and supervision
both when the engines are first started and after any over-
hauling. It is a good plan to bar the engine round before
turning on the steam whenever the engine has been stopped
for repairs. The luxury of a barring engine is enjoyed by
few colliery engineers, although it is a great convenience,
especially in large rope-driving fan engines and air compressors having Corliss valves. It is not only useful for
starting purposes, but it is of great assistance when putting
new ropes on the pulley, and of special advantage when
setting or adjusting the valves.

When taking off the cylinder or valve-chest cover of an
engine that has just stopped, and may still have uncondensed
steam in the valve chest and cylinder, the nuts of all the
studs or bolts should be eased off equally, then a wedge
or chisel inserted to break the joint, unless forcing screws are
provided, as is usual in most engines. This allows whatever
vapour remains in the cylinder or valve chest to escape;
whereas, if each nut were taken off its thread singly, it
might happen that the joint would be broken by the pressure
of the steam when all but a few nuts were removed. The
probable result would be that the remaining nuts would be
wrenched off, or the attendant injured. In the case of a
cylinder the indicator cocks would, of course, be first opened,
and the foregoing precaution would scarcely be necessary,
but there are some valve chests where it is not easy to allow
the enclosed steam to escape. Some may think many of the

above remarks appear superfluous, but their neglect certainly
produces chance of mishap. The most experienced
engineer must have been quite inexperienced at some portion
of his career, and it is better some part of his experience,

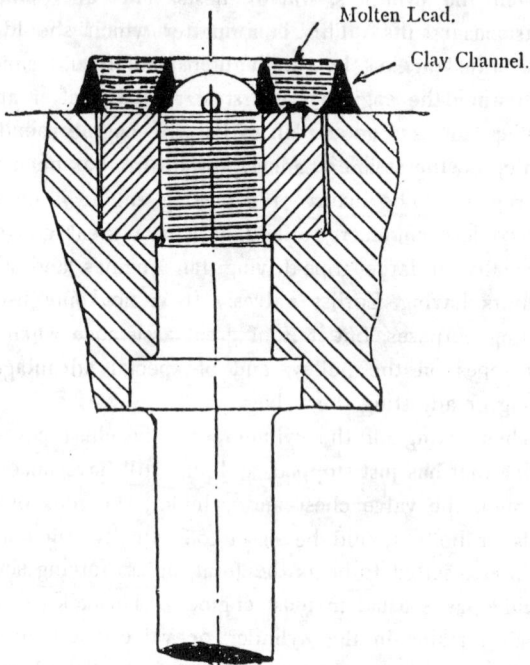

Fig. 7.

at any rate, can be gained by other means than actual breakdowns.

When dismantling an engine for repairs or removal, great
trouble is often experienced in getting various parts asunder.
This is chiefly due to corrosion and a result of not well
greasing the parts when put together. The piston nuts are

generally the worst to get off, especially when they are recessed in the piston. When the nut is external a good plan is to encircle a ring of red-hot iron round it for a few minutes. This heats the nut and causes it to expand slightly before the heat penetrates to the rod. Then remove the ring and apply the spanner as quickly as possible. When a crab or winch is handy the end of the spanner can be hitched on to the cable, and a good purchase thus obtained. Holding down bolt nuts, valve spindle nuts, and collars of various descriptions can often be removed by the above method when mere pulling at a cold nut would be of no avail. In cases where the nut is recessed in the piston there is probably no room to

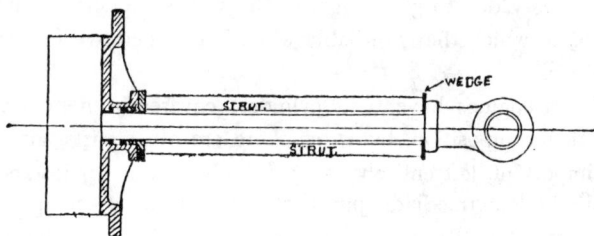

Fig. 8.

get a hot ring between the nut and the recess. This renders the operation of removing a fast nut very tedious. The following method, although somewhat elaborate, is usually successful. Tilt the piston and rod on end as shown in sketch (Fig. 7), and on the nut form a clay channel, as illustrated. Also lightly cover the tommy holes or spaces for the key. Now pour hot lead into the channel and when firmly set knock the clay and lead away, and apply the spanner. Sometimes the piston and crosshead cannot be got off the piston rod by ordinary means: especially is this the case with air-pump rods, where corrosion is rapid. Here again the expansion of metals proves a valuable property and can be employed to force the parts asunder. The accompanying figure (8) is almost self-explanatory. The crosshead is the piece to be forced off. The rod is first heated by a fire. This is then quickly removed and stout pieces of iron are placed between the crosshead and any convenient immovable part on the rod or in connection with it, and wedged up tightly. On cooling, the crosshead will be eased off or the struts will bend by reason of the contraction of the piston rod when cooling. Something is bound to give. The engineer should satisfy himself that the part receiving the thrust of the struts is quite strong enough, and not subject to deflection, or the operation will prove ineffectual.

All escape valves and the like should be lifted off their seats every day to guard against the valves sticking on their seats, to which they are liable when left inoperative for long periods.

Most engines have their cylinders covered with non-conducting composition, but often there are other parts, almost as important, left entirely bare; cylinder covers for instance, and the intermediate pipes and receivers of compound engines. Small economy is to be expected unless all such places are well clothed.

There are several points which require attention when an engine is stopped for any lengthy period, say exceeding a fortnight. The cylinders, pipes and valve chests should be thoroughly cleared of water; the packings, whether metallic or otherwise, should be removed from all glands, and the rubber valves in the air pump should be removed, cleaned, and carefully stored. All the bright parts may be thickly coated with grease, and every day the engine should be barred through a portion of a revolution, so that on starting again it will move more freely than it would otherwise have done, and there is less chance of scoring the wearing surfaces and straining the motion work. All the wicks and worsteds should be removed from the lubricators, the journals simply receiving a copious feed of oil from a can soon after stopping.

It is not an uncommon occurrence for a crack to appear in the bedplate, cylinder feet, and cylinder and valve-chest covers. It may be a mere surface crack, and of little consequence, but it is the business of the engineer to discover its nature, cause, extent, and whether it is spreading. The bedplates of air compressors and direct-acting pumps are particularly liable to fracture. No class of machinery about a colliery is more severely stressed. At the end of the stroke there is the full pressure of air or water on the piston, and at the same time, if there be lead on the steam cylinders, there is the full pressure of steam acting on the steam piston in the same direction. The result is that a violent blow is brought on the engine at the end of every stroke, which will soon discover any weak spot. Air compressors and pumps should therefore have no lead whatever, and very little compression. But to return to the previous subject: when first discovered, the surface of the metal round the crack should be wiped clean, and petroleum painted thereon. The petroleum should then be wiped off, when the length of the crack will be more defined. Now make a mark with a centre punch at each extremity of the crack. This will at once show whether the crack is extending and at what rate. The petroleum will continue to ooze from the crack for some days after painting, thus rendering the extent of the crack more apparent.

When packing the pump glands of pumping engines the following point is worth attention. The packing usually consists of square asbestos cut into lengths suitable for encircling the pump plunger. In fitting this packing it is advisable to let the ends be short of each other by an amount equal to the thickness of the packing. Thus, if the packing be ½-inch the ends should be ½-inch apart when wound on the plunger. This allows room for the swelling of the packing which takes place when it becomes saturated with water. If no clearance were allowed the packing would become unduly tight. This is worse than a little weeping at the glands so far as economical working is concerned. It is possible to bring a plunger pump to a standstill by unduly tightening the glands.

Should the injection water fail a leakage at one of the joints, or fracture in the range, may be the cause; but more frequently the failure is due to the pipes becoming choked up, especially if no sieve or strainer box is fitted to the lodge end of the range. Dead dogs and cats are common wild fowl in some lodges, particularly in Lancashire, and amongst other peculiarities they seem to have a liking for entering suction pipes. Strainer boxes curb their impetuosity and prevent them entering except in harmless instalments. Air pumps are often very unsatisfactory in working, and through being placed in dark inconvenient holes they do not invite attention.

Engines that must run for long periods and without lengthy stoppages are usually fitted with spare parts; and the following is a list of the gear that is usually supplied in duplicate:—One set of crank pin brasses, one set of crosshead pin brasses, two connecting-rod bolts (if of marine type), one complete set of piston rings and packing for each size of cylinder, spare set of springs for relief valves (one for each size), one set of rubber valves for air pump delivery grid, one rubber foot valve, one set of rubber valves for air pump bucket, and several spare piston junk ring screws. Corliss engines, in addition to the above spares, should be provided with :—One set of dashpot springs, one set of tripper springs,

one set of hardened steel catches or bits, and one complete
set of cushion leathers (if any).

The above list applies to an ordinary condensing engine,
either compound or otherwise, and for many engines the
list would be supplemented. Air compressors, for instance,
should have spare inlet and outlet valves and springs; and
pumps should have spare valves and seats. Important
hauling engines may have spare pinions and spur wheels, and
rope-driving fan engines several spare ropes.

CHAPTER V.

INDICATOR DIAGRAMS.

THE indicating of steam engines and the interpretation of the meaning of various kinds of diagrams is one of the most interesting subjects connected with the steam engine. Every engineer ought to understand how diagrams are taken and be able to calculate the power of his engine from them. Besides giving a clue to the power the engines are giving out they discover many defects which would otherwise escape notice. They show whether the valves are properly set, whether the ports are too small, whether serious leakages are present, and also whether quieter running can be obtained by adjusting the compression; and in large engines they show whether the friction of the engine is excessive.

The essentials of an indicator are an accurately fitting piston loaded with a spring of known strength, and an oscillating drum on which a paper can be fixed. The piston of the indicator is under the same influences as the engine piston which is being indicated, and the drum is connected by a string through suitable mechanism to the piston rod or crosshead, and the motion of the string is an exact reproduction on a reduced scale of the actual motion of the piston. The direction of the indicator piston motor is at right angles to the motion of the drum.

The drawing (Fig. 9) shows the section of a Richard's indicator, which for speeds not exceeding 120 revolutions

per minute is mostly used, and is unsurpassed. The piston
rod A is connected at the top to a parallel motion carried
by the arms B. These arms are cast together and swivel
round the cylinder body. The pencil point C is attached to

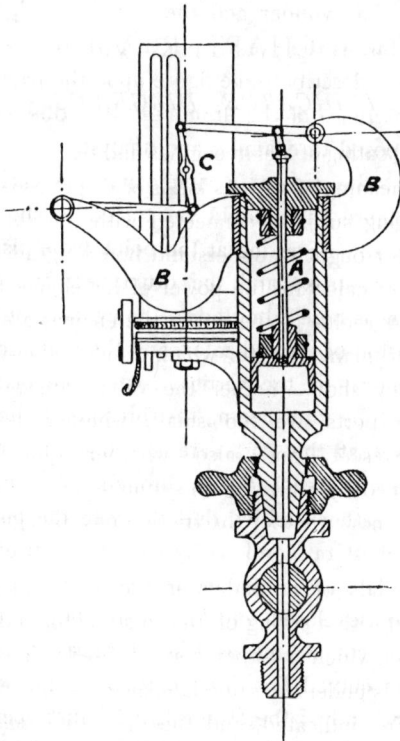

Fig. 9.

the parallel motion and by moving the arms B the point
of the pencil can be made to press upon the surface of the
moving drum. At the base of the drum there is a coiled
spring which insures that the drum gives a tension on the string
at all times and performs the return movement of the drum.

It will now be understood that if the string be connected to some mechanism which has a similar motion to the piston rod, but on a reduced scale, so that the drum moves something less than a complete revolution, and the indicator screwed into the cylinder and the cock opened, the piston will rise and fall as the pressure in the cylinder varies. The pencil will then describe some figure upon the paper fastened to the circumference of the drum, and it is this figure which it is now proposed to examine and analyse.

Taking first the case of a simple non-condensing engine, a probable figure that would be traced is shown by the diagram (Fig. 10). Starting at the point A the piston is near the end of the stroke and the drum near the end of its travel. Steam is here admitted to the cylinder of the engine and to the piston of the indicator, and the pressure rises very rapidly, as shown by the vertical line. The piston is now urged forward by the steam, the pressure falling a little, which shows that the ports are not big enough to allow the full amount of steam to pass through. At B the pressure begins to fall rather rapidly, due partly to the piston travelling at its highest rate and partly to the action of the slide-valve which is now beginning to close the port. At C the port is quite closed and the steam begins to expand. Expansion, of course, means loss of pressure, and this is shown by the curved line from C to D. Before the piston is quite at the end of the stroke the exhaust opens and the pressure rapidly drops as shown, and at the end of the stroke the pressure of steam is represented by the height EF. The indicator pencil now traces an approximately horizontal line as the engine's piston makes its return stroke, but towards the end of the stroke the valve closes for compression, and the steam being confined in the port becomes compressed by the advancing piston. This causes a rise of pressure in the cylinder as indicated by the line on the diagram G to H. At H the valve opens the steam port and the pressure at once rises and the pencil begins to trace another diagram.

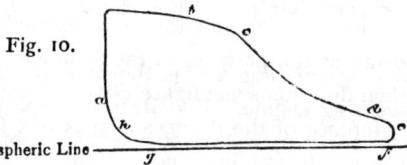

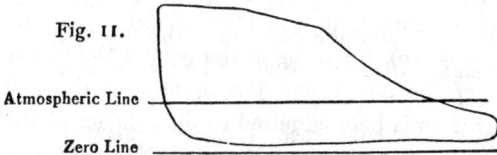

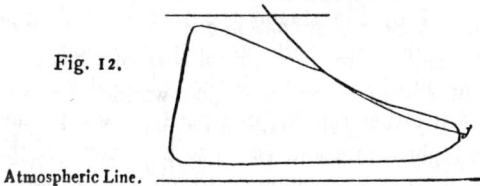

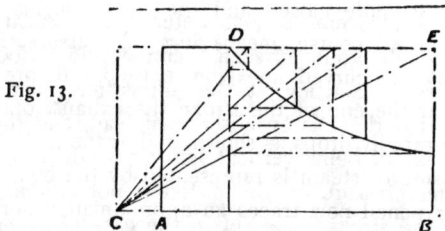

Fig. 10.

Atmospheric Line

Fig. 11.

Atmospheric Line

Zero Line

Fig. 12.

Atmospheric Line.

Fig. 13.

The exhaust line is slightly above the atmospheric line which shows that the exhaust is not perfectly free. The atmospheric line is traced by the pencil of the indicator when the cock is closed and the drum is oscillating under the influence

D

of the string, the pencil being held lightly against the paper.
If the string were disconnected and the indicator taps opened
the pencil would simply describe a vertical line which, beyond
recording the maximum pressure in the cylinder, would have
no value.

The next diagram (Fig. 11) is the type of card obtained
from a condensing engine. It is in most respects similar to
the previous diagram, but a considerable portion of the
figure is below the atmospheric line, and, as in the diagram
from the non-condensing engine, distances above the atmospheric line indicated the pressure of the steam, so in the condensing engine diagram distances below the atmospheric line
show the degree of vacuum in the cylinder due to the action of
the air pump and condenser. The line of no pressure or perfect
vacuum has been drawn to scale on the figure, and the closer
the exhaust line approaches to it the more efficient is the
condenser. A close examination and study of these diagrams
will enable any one to form a clear idea of what takes place
in the cylinder of a steam engine working under normal
conditions, but the most interesting part of the subject lies
in the consideration of unhealthy diagrams, and the best
means of curing the diseases. Taking the case represented
by Fig. 12, several serious defects are indicated, and an
engine which produced such a card would be both wasteful
and noisy. Starting at the admission point, it will be
noticed that there is an absence of lead, and the admission
line instead of being vertical slopes considerably, so that the
maximum pressure on the piston is not attained until it is
well on the stroke. Except in the case of a compressor or
pump, this is a bad state of things. The next defect to
notice is the sloping of the admission line, which shows that
the steam is throttled somewhere; it may be in the pipes
from the boiler past the stop valve, past the valve, or in

the ports. The diagram of course does not indicate where, but it will be shown later how to discover the place where this throttling or wire drawing takes place. Throttling or wire drawing is not in itself a defect, except that when an engine is working with a full load with much throttling it is necessary to admit the steam later in the stroke than it would otherwise be, and consequently the expansion is curtailed and the economy thereby reduced. The most serious defect in the figure is the expansion line, but before considering this point the method of drawing the theoretical hyperbolic curve should be understood.

Let AB (Fig. 13) represent the stroke of a piston, and AC the clearance volume. Let the point of cut-off be at D (quarter-stroke in diagram). Draw the diagonal CE and draw any number of lines from C to points in the line between D and E where these points cross the vertical and draw horizontal lines to intersect verticals drawn from the points at which the diagonal lines intersect the line DE, then draw a curve through these points. This is known as the hyperbolic curve, and represents graphically the behaviour of a gas expanding according to Boyle's law, that is, the gas is at a constant temperature throughout the expansion; and at all times the pressure multiplied by volume is a constant quantity. No steam engine exists in which this condition is realised, but it is a convenient standard for comparing the actual expansion curves of indicator diagrams.

Applying this curve to the diagram under consideration, it is found that the actual expansion line is outside the theoretical curve. Now it is impossible for a non-jacketed cylinder using saturated steam and having tight valves to produce such a result, and the explanation of this rise above the theoretical curve is a leaky valve. Although the leakage is only apparent in the expansion part of the diagram, ·it is probable that it is going on throughout the diagram, and may in part account for another defect to be noticed. At J, almost at the end of the stroke, the exhaust opens, but instead of the pressure falling rapidly it is 15 per cent. in the return stroke before it becomes horizontal to the atmospheric line. This shows that the exhaust is not allowed to escape soon enough, and the engine is choked. The exhaust of such engines is generally a long-drawn rush, instead of a clear defined puff. The exhaust line never approaches within 6 lbs. of the

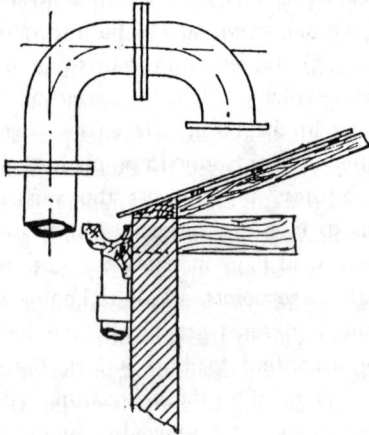

Fig. 14.

atmospheric line. This is a poor result and would seriously reduce the power of the engine. The leaky valve might account for some of this back pressure, but the probability is that the exhaust pipe has been choked up. Sometimes slates and portions of chimney pots get into the pipe when they are arranged with a vertical open top. When the exhaust pipe ends near the roof of a building it is therefore a good plan to put in a couple of elbows and arrange the opening somewhat as shown in Fig. 14.

The last defect is the absence of compression. This is wasteful because the next admission of steam has to fill up the clearance space, whereas if compression had occurred the ports would have been already filled with compressed steam. In addition to this, the working of the engine would be noisy because there would be no check for the momentum of the moving parts.

Besides the defects in diagrams dwelt upon in the previous chapter there are a few others which may arise through no fault in the engine but through the defective construction or bad condition of the indicator. It is important that these should be discriminated from defects in the steam distribution, otherwise the engineer may be seeking to remedy evils which do not exist. A loop at the admission line of diagrams may be caused by excessive compression or by the presence of water in the cylinders, but very often the real cause is the jumping of the indicator pencil. When the steam is admitted to the cylinder the indicator piston at once flies up, and with light springs the momentum will carry it beyond the point due to the pressure of the steam alone. This excess is followed by a vibration in the other direction, and as the paper is moving to and fro during the oscillation a loop is formed. More commonly a series of peaks is formed. The irregularities will usually present themselves on high-pressure engines, especially where a weak spring has been chosen.

A diagram about four inches long and two inches high is a convenient size; therefore if the boiler pressure were 100 lbs. a spring of 50 lbs. to the inch would be the best to employ. The travel of the spring is practically equal to the length of the diagram, because the pulley and the drum are about the same diameter; so that with an engine three-feet stroke it would be necessary to use indicating gear that

would reduce the motion of the piston to one-ninth in order
to obtain a card four inches long.

When engines are provided with expansion gear and the
point of cut-off is under the control of the governor, unless
diagrams from both back and front ends of the cylinders are
taken simultaneously there may be an inequality in the
diagrams which might lead to the conclusion that the valves
were unequally set, whereas it will be probable that this
is not the case. An example of this kind is shown by the

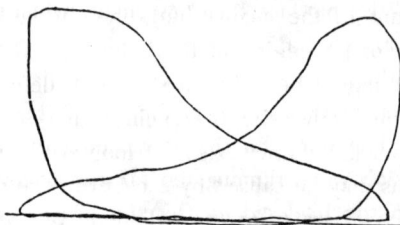

Fig. 15.

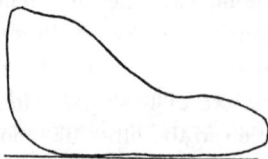

Fig. 16.

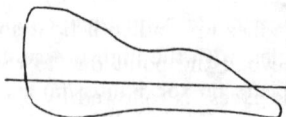

Fig. 17.

diagrams in Fig. 15, taken from an engine of the writer's
design. The valve gear is of the Corliss type, and in the
high-pressure cylinder from which these diagrams were
taken the point of cut-off is regulated by the governor.
The instant at which the valve closes differs greatly
in the two cards and is accounted for not by inequality
of valve setting but because the load in the mill varied in
the interval of time between the taking of the figures, and
the governor adjusted the cut-off to suit, thus keeping the
speed uniform. The engine in question is a tandem horizon-

tal compound condensing, high-pressure cylinder sixteen inches diameter, low pressure thirty-two inches diameter, three feet six inches stroke, revolutions seventy-five per minute. The indicated horse-power is 240. It is driving cotton spinning machinery at the Brunswick Mills Co.'s factory at New Mills, Derbyshire. Sometimes a series of steps or waves appear on the expansion curve. These are usually accounted for by grit or water in the indicator or looseness of the joints in the parallel motion. Unskilful use of the indicator may also produce them, especially when the pencil is pressed too tightly to the paper.

Engines with slide expansion gear are liable to give diagrams similar to that shown in Fig. 16, if the valve gear is defective. The lump on the expansion curve is caused by the expansion valve running too far and uncovering the steam port on the back edge after the proper cut-off has taken place. Such a state of things is very objectionable and should at once be remedied. In the low pressure diagrams of twin compound engines with cranks at right angles, with the high pressure leading, a peculiar hump often appears on the admission line. Fig. 17 is the low pressure diagram from a large compound engine, and shows this peculiarity. It is not to be supposed that inefficiency of the valve gear has anything to do with this. The cylinder draws its supply of steam from the pipes leading from the high-pressure cylinder. As the capacity of these pipes together with that of the steam chest is not in excess of the cylinder capacity the pressure falls considerably as the piston advances on its stroke, but shortly before half stroke the high pressure exhaust port opens and a fresh supply enters the pipes, thus causing a rise of pressure before cut-off in the second cylinder takes place.

In order to calculate the power from diagrams it is neces-

sary to find the average pressure. This is usually found by
drawing ten vertical equidistant lines, measuring the pressure
on each line, and taking the average of the ten separate
readings. When any loops appear on the figure they must
be deducted from the average pressure, because a loop
shows that an engine is doing negative work, or as it might
be expressed the piston works the steam, and not the steam
the piston. The diagram shown by Fig. 18 shows the method
of working out the horse-power, and also serves to illustrate
the effect of a loop on the calculations. In this case the
engine is much underloaded, or in other words, it is too large
for the work it has to do. The cut-off occurs very early in
the stroke and as expansion is carried on the pressure falls
below that of the atmosphere. Now the pressure on the
other side must be at least equal to atmospheric pressure,
and therefore instead of an impulse on the piston there
is retardation and the piston must be dragged along by the
fly-wheel. The loop indicates the magnitude of the drag
and is therefore to be subtracted from the rest of the
diagrams in order to find the actual indicated horse-power.
The diagram serves as a specimen card showing the data
which ought to be recorded on every figure, or series of
figures. It will be assumed that having these particulars the
reader is able to work out the horse-power himself. The
area of the piston rods ought to be taken into account as
with high pressure it is responsible for a considerable reduction of power. To illustrate this take the case of an engine
having a piston rod at both back and front five inches
diameter, piston speed 500 feet per minute, and a mean
pressure of 40 lbs. The reduction of power due to the rod
in this case is $\frac{19\cdot6 \times 500 \times 40}{33000} = 12$ horse-power nearly.

In the next chapter practical instructions for taking

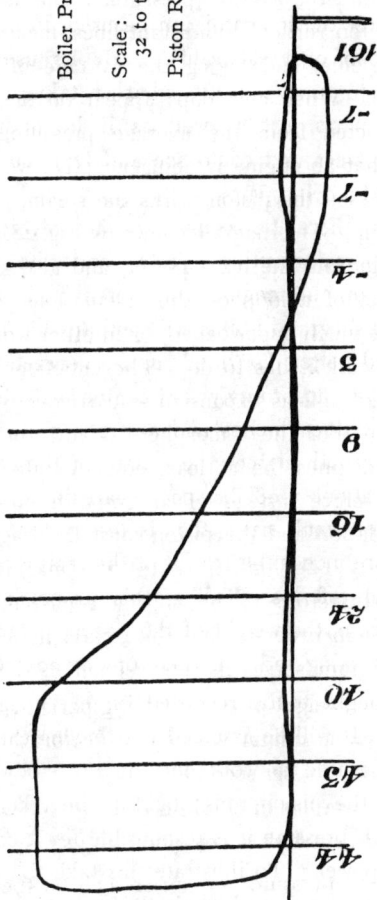

Fig. 18.

Name _____

Date _____

Boiler Press.: 60 lbs.

Scale: 32 to the inch.

Piston Rod: 3 in. dia.

Net total: 161

I.H.P.: 43·9
Average Pressure: 16·1 lbs.

Front end

Cylinder 18 in. dia.

Stroke: 3 ft.

Revolutions: 60.

accurate diagrams will be given, and it will afterwards be
shown how the indicator can be used, in many places giving
information of great value.   By its use the engineer is
enabled to know exactly what is taking place in all the
internal parts just as well as if he actually entered the
various places and could see and feel the behaviour of the
fluids and moving parts.

It will here be convenient to revert to the inertia
of the reciprocating parts and its effect on the
working of the engine considered with reference to the
indicator diagram.   This will perhaps be more easily understood and better described from a concrete case rather than
abstractly; accordingly an engine will be taken with a
cylinder sixteen inches and a stroke of twenty-eight inches.
For horizontal engines the weight of the reciprocating parts
is often about $3 \cdot 98$ lbs. per square inch of piston area for
engines of small size; that is to say that with a piston of
100 square inches the weight of the reciprocating parts would
be about 398 lbs.   Taking this proportion, then, for the
present case, the weight of the piston, piston rod, crosshead,
and connecting rod will therefore be $201 \cdot 06 \times 3 \cdot 98 = 800$ lbs.

Let W = weight of reciprocating parts in lbs.,
  V = maximum velocity of reciprocating parts in feet
      per second,
  L = radius of crank in feet,
  G = gravity's acceleration = $32 \cdot 2$.

Then the pressure per square inch on the piston necessary
to start the mass of the parts =

$$\frac{W \times V^2}{G \times L \times \text{area of cylinder.}}$$

The length of the connecting rod introduces a small error
in the calculation, but as this is only slight it may here be

neglected. Applying the above formula to the present case
and taking a speed of 100 revolutions per minute the equa-
tion becomes :—

Pressure per sq. in. necessary to start or stop moving parts

$$= \frac{800 \times 12 \cdot 2^2}{32 \cdot 2 \times \frac{7}{8} \times 201 \cdot 06} = 16 \text{ lbs. nearly.}$$

This means that if the total pressure on the piston at
the beginning of the stroke were 100 lbs. only 84 lbs. would
be available for urging forward the crank pin at the beginning of the stroke, 16 lbs. being absorbed as momentum in
the moving parts. Now, although this force of 16 lbs. on the
square inch is thus diverted from its intended effect at the
beginning of the stroke, exactly the same amount (neglecting
friction) is given out to the crank pin in the second half of
the stroke; and at the end of the stroke the pressure on the
crank pin would be the same as that produced by 16 lbs.
acting on each square inch of piston area if the engine were
scotched at the fly-wheel, and in the case under consideration
unless the pressure on the reverse side of the piston were
equal to 16 lbs. the reciprocating parts would hurl themselves upon the crank pin, and if the brasses were at all slack
a knock would occur; and in any case the crank pin would
be unnecessarily stressed : whereas if compression of the
exhaust to 16 lbs. took place the engine would turn the
centres without undue strain or stress. The action of inertia
can be very clearly shown on the indicator diagram as in
Fig. 19. To the scale of the diagram AB is marked off equal
to lbs. pressure, and CD = AB. A line joining B and D shows
graphically the action and degree of inertia at any point in
the stroke. In the first half of the stroke it is against the
steam pressure and the resulting thrust on the crank pin is
the *difference* between the steam pressure and the inertia
pressure taken at any point. Thus at 25 per cent.

in the stroke the steam pressure is represented by EF, the pressure due to inertia is EG, and the actual thrust transmitted to the crank pin is FG. In the second half of the stroke the resulting thrust on the crank pin is the *sum* of the two pressures. From this it will be seen that the thrust on the crank pin is very different from what might be inferred from the indicator diagram, and in expansion engines is considerably more uniform because of the effect of the moving parts. In well-designed engines the action is exceedingly

Fig. 19.

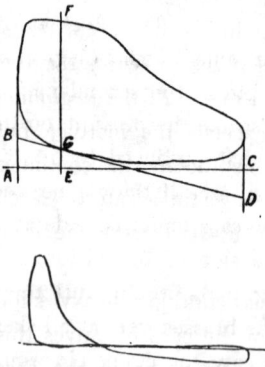

Fig. 20.

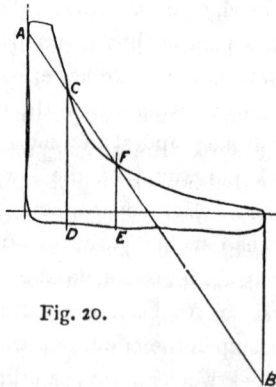

Fig. 21.

harmonious. At the beginning of the stroke, when the greatest force comes on the piston suddenly, the reciprocating parts absorb much of the thrust, but in the second half of the stroke, when the pressure in the cylinder begins to fall by reason of the expansion, this stored energy is given out just when it is most desired. It will be noticed that the compression curve is quite above the line BD, which indicates that compression completely absorbs the inertia thrust.

It were an interesting calculation, though of no practical

value, to estimate the speed at which any given engine would reach self-destruction, supposing the fly-wheel and valve gear were good for any speed. It is also interesting to calculate, with a given compression, the speed at which the engine would be liable to knock, and at what point. It would explain that well-known phenomenon in the running of engines, that up to a certain speed everything works smoothly, but when a certain speed is exceeded a great clatter is set up.

Besides a reversal of stress at the end of the stroke it is possible to have reversals at intermediate points. This may occur in high speed engines having a high ratio of expansion in the cylinder. Let Fig. 20 represent the indicator diagram from some engine, and AB the inertia line. At the beginning of the stroke the steam pressure exceeds the inertia; but at CD the two are equal. After this the inertia continues in excess to EF, when the pressure is again greatest. Such an engine would knock at C and F if there were the least play in the crank pin brasses.

The pins and rods of valve gear working heavy slide valves are stressed by inertia, and in the case of valves this inertia cannot be checked by compression as in the case of a piston and connecting rod; hence the severe and rapid wear of valve gear beyond the degree that the resistance to motion arising from steam pressure and friction would seem to cause.

The question of inertia is an important one for an engineer, therefore it has been dealt with at some length. It will enable the engineer to understand the stresses set up in an engine more clearly than the mere indicator diagram would, and will show why some engines, however carefully adjusted, soon become noisy, whilst others work quietly even when the brasses become badly worn.

There are two methods of finding the internal friction of an engine, that is, the power required to move itself. The most exact method, but the one which is impracticable for large engines, is to find the brake horse-power when working with a certain fixed load, and running at the normal speed, indicator diagrams at the same time being taken from all the cylinders. The difference between the brake and the indicated horse-power represents the power absorbed in moving the working parts of the engine. The other method is to run the engine at full working speed without load, and a diagram taken under such conditions will give, approximately, the power lost in friction. Friction diagrams have a very attenuated appearance, often with loops at the end of the expansion curve. Such a diagram is shown in Fig 21. It is from the cylinder of a Corliss engine, which under normal conditions indicates 286 horse-power. The indicated horse-power represented by the friction diagram is fourteen, thus giving a mechanical efficiency of 95 per cent. nearly.

A few remarks about friction brakes and the method of measuring the brake-power of engines will here be convenient.

Fig. 22 represents a common form of friction brake, and will serve to show the method of application as well as the manner of estimating the power given off. Referring to the figure, a series of hard wood blocks are curved to fit the rim of the fly-wheel or driving pulley, and strung together on a continuous double hemp rope. To the dragging end of the rope a spring balance is attached, whilst the other end terminates at a hook to which either weights or a spring balance may be hung. The method of testing is as follows: —The brake being in position, but only a slight weight or tension being put on the rope, the engine is started and gradually put on to full speed. The tension on the hook end of the rope is then increased, but the stop valve being opened wider the engine is still kept up to its speed. The increase in weight and opening of the stop valve is continued until the maximum load the engine will support at the full speed is reached, when the pull on the spring balance is read off. The horizontal distance from the centre of the crank shaft to the centre of the balance being known (see distance marked R in Fig. 22) and the speed of the engine at

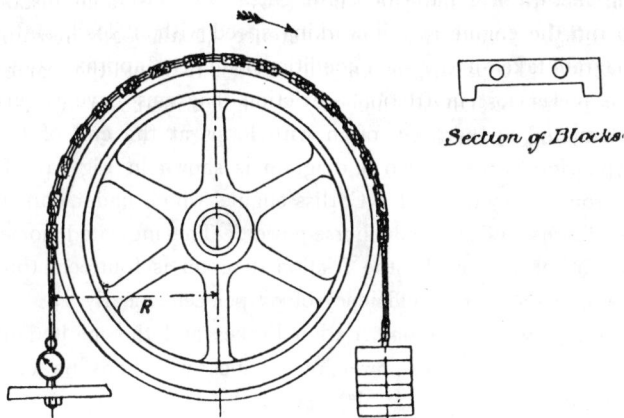

Section of Blocks

Fig. 22.

the reading of the balance, all necessary data are to hand for the calculation of the horse-power.

Let R = distance from centre of shaft to centre of balance
      (see Fig. 22).

W = pull on the balance in lbs.

N = number of revolutions of engine per minute.

Then the brake horse-power $= \dfrac{2\,R\,\pi \times N \times W}{33{,}000}$

This formula will be best understood and remembered when it is considered that it estimates the number of linear feet per minute slipping past the brake, referred to the radius at which the balance acts. This number of feet moved through multiplied by the pull gives the work done per minute in foot-pounds, which divided by 33,000 gives the horse-power.

The above described method gives the maximum power the engine is capable of developing, but of course it is possible to measure the power at part 'oads. When compared with the indicated horse-power at the same instant it will be found that the difference at all loads remains practically constant, when the speed remains the same; hence the mechanical efficiency of engines increases in direct proportion to the load.

CHAPTER VI.

TAKING INDICATOR DIAGRAMS.

HE precautions to be observed in fixing up temporary indicating tackle are that the joints should be without play and the mechanism arranged to give a true motion to the string. The simplest indicating gear consists of a board swinging on a centre carried above the slide bar as shown in Fig. 23. A slot is provided at the lower end to

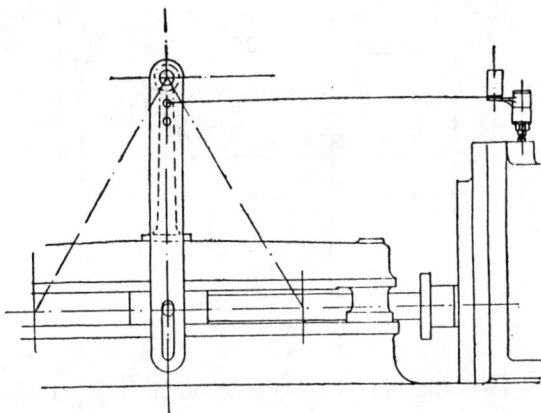

Fig. 23.

pass a small stud fastened in some convenient place on the crosshead or slide block. At some point on the board having a movement of 4 inches or thereabouts a peg is inserted to which a string is attached. Although this does not give a true motion to the string it is quite good enough for ordinary cases, provided that the distance from the centre of the

E

stud on the crosshead to the centre of the carrier shaft is
not less than the stroke of the engine.   The string should
be led off at right angles to the board when vertical.   This
is important, and if neglected a very distorted diagram will
result.   By placing a grooved pulley on the counter shaft
and attaching the string to it the latter may be led off at
any angle without any difference to the motion, though the
error arising from the varying position of the stud in the slot
is still present.

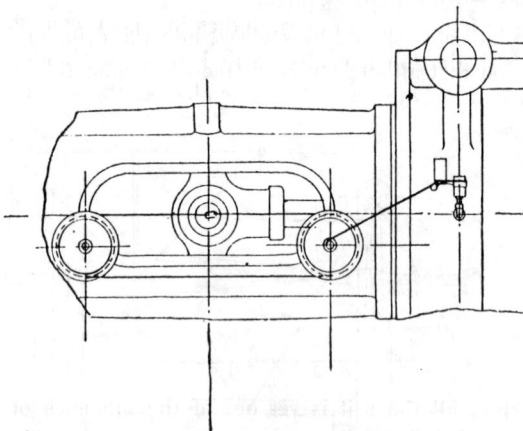

Fig. 24.

Sometimes the indicating gear shown in Fig. 24 is used.
This form gives to the string an exact reproduction of the
piston's motion on a reduced scale, but if not carefully made
and if used on high-speed engines it will be liable to error.
At each end of the slide bars two light wheels are mounted
and a string passes from a stud on the slide block round
each pulley.   The indicator string is fastened to the hub of
one of the wheels, and it is immaterial at what angle the
cord is led.   As the crosshead moves the wheels revolve

backwards and forwards, and unless they are of light construction, the string tight in the grooves, and the speed slow,
they are apt to overrun the string at each end of the stroke
and distort the diagram. The pantagraph motion shown in
Fig. 25 gives an exact motion to the string and is suitable
for all speeds, but the string must be led off in a direction
parallel to the piston rod and carried to the desired point
by one or more small guide pulleys. The gear is shown
in mid-position, the dotted lines indicating the position it
assumes at each limit of travel.

The indicator should be screwed into the cylinder in such
a position that although it is in free communication with the

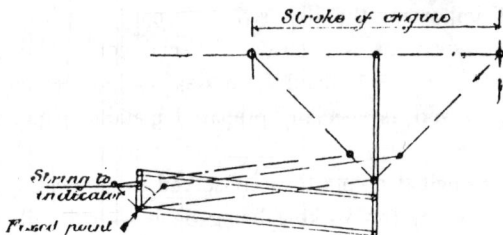

Fig. 25.

cylinder at all times it is yet out of the influence of steam
currents. The best place for attachment is on the cylinder
cover on the side remote from the steam port entrance, but
such a position is not convenient for the string, especially at
the back end. The most usual place for the indicator connection is on the top of the cylinder between the piston and
the cover. It should be known however, by actual trial,
that when at the end of the stroke the piston does not cover
the indicator hole, otherwise misleading diagrams will result.
Indicators should never be fixed on the steam ports because
the rush of steam backwards and forwards prevents the
exact state of things in the cylinder being recorded on the

card. It is very convenient to be able to take diagrams from each end of the cylinder without removing the indicator, and this is readily accomplished by taking a pipe from each end of the cylinder and connecting the indicator to a three-way cock in the centre; but this arrangement, though convenient, is open to objections. In the first place the condensation in the pipes must affect the figure to some extent, and even if they are well clothed the true state of things in the cylinder cannot be recorded at the indicator through several feet of small piping. The only satisfactory way is to employ two indicators; although this is rarely done except the engines are undergoing special tests and trial runs.

It is important that the string or cord is not elastic, or serious distortions will ensue. Special cord for indicating purposes can be obtained from dealers in scientific instruments, as well as specially prepared metallic paper for the drum.

The principal points to be observed in connecting up the indicator ready for working being dealt with, it will be convenient to give a few hints on the actual taking of diagrams. Having connected the string to the drum, the engine running meanwhile, lightly press the pencil upon the paper: the straight line thus drawn is the atmospheric line. Now swing the pencil back from the paper and open the indicator tap. Let the indicator work for a few seconds before again bringing the pencil to the paper. This allows the indicator to get warm and the piston and motion to work freely. Then bring the pencil to the paper, when the diagram will be drawn. The pencil should not be held to the paper longer than is necessary to give a complete diagram, otherwise the figure becomes indistinct. When the diagram has been drawn swing back the pencil and shut the indicator tap, then uncouple the string and remove the card from the drum. The following particulars should be noted on every diagram taken :—

Date and time of taking.
Scale of spring.
Boiler pressure.
Number of revolutions per minute.
Vacuum as read from gauge.
Diameter of the cylinder.
Length of the stroke.
Diameter of the piston rod at both back and front end.
Any special circumstances influencing the working of the engine.

Indicators should be very carefully kept; the cylinder should be well dried and wiped clean after use, and afterwards treated with the best sperm oil, the piston and motion work also receiving similiar attention. The chief considerations which determine the choice of the spring have been mentioned in the previous chapter. If too weak a spring were chosen the indicator piston would strike the cover and the resulting diagram would not be a true record of what had taken place in the cylinder. On the other hand, too stiff a spring gives a small diagram and slight peculiarities in the figure become indistinct. It is a curious circumstance that no two engines give diagrams exactly alike. You may take two engines built from the same patterns and apparently identical, set them to do the same work and (as far as can be observed) under the same conditions, yet the diagrams will differ. An individuality seems to be present in every engine, as though the moods and feelings of the men who made them impressed a certain subtle influence and imparted a distinct temperament on each. Many engineers are in the habit of personifying their engines, and this is no mere slang, but the acknowledgment that engines have, if not a consciousness,

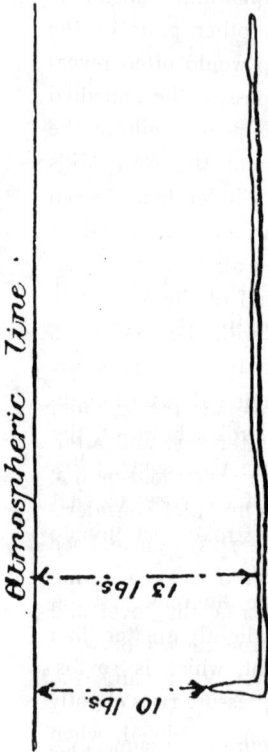

Diagram from the front end of the pump. Scale 1/10th.

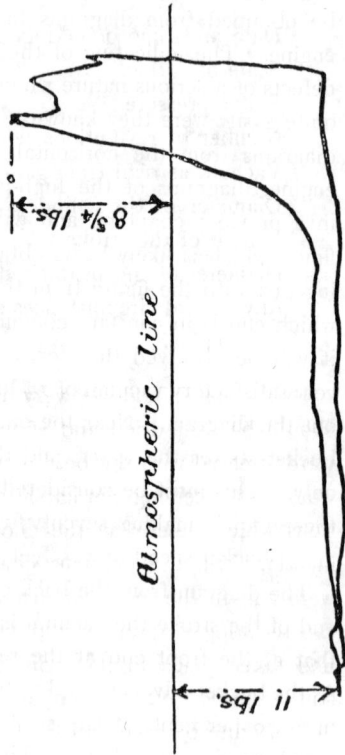

Diagram from the back end of the pump. Scale 1/10th.

Fig. 26.

at least idiosyncrasies and moods which are supposed by most people to be the attributes of living things alone.

Although the use of the indicator is chiefly confined to engine cylinders, information of almost equal importance is to be obtained from diagrams taken from other parts of the engine. The indicating of the air pump would often reveal defects of a serious nature which might possibly be remedied quite easily were they known. Fig. 26 is a fac-simile of the diagrams from the horizontal air pump of the New Mills engine, diagrams of the high-pressure cylinder being given in a previous chapter; and will serve to show the kind of figure which is likely to be obtained from air pumps. Dealing first with the figure from the front end of the air pump, which end is in constant communication with the condenser, it will be observed that for a horizontal type of pump the very satisfactory vacuum of 13 lbs. is obtained almost throughout the diagram. Near the end of the stroke, however, the bucket strikes the water, and the pressure rises some 3 lbs. only. This must be considered very satisfactory, seeing that the pump is making seventy-five double strokes per minute and a bucket speed of 525 feet per minute.

The diagram from the back end is quite different. At the end of the stroke the vacuum is 11 lbs., slightly greater than that at the front end at the same instant, which is 10 lbs. As the bucket advances in the stroke the pressure rises slightly until 70 per cent. of the stroke has been completed, when a sudden rise commences. This shows that the bucket has met the water and is forcing it through the delivery grids into the hot well. The pressure rises to 8¾ lbs. above atmosphere, which is somewhat of a shock and shows that the area of the delivery valves might with advantage have been greater. The figures are not very intelligible without a drawing of the air pump, accordingly the accompanying diagrammatic sketch (Fig. 27) is presented, which will illustrate the main features. The bucket, it will be noticed, is without

valves, the water and air entering the air pump barrel as the
bucket uncovers the ports at the end.   On the return this
water and air are trapped between the bucket and delivery
grid and towards the end of the stroke are forced through
the valves.   The points at which the diagrams were taken are
shown at xx; the overflow is at B, and the feed-water is taken
from c.

   When an indicator diagram shows a rapidly sloping admission line it is not always to be concluded that the valve gear
is at fault.   The stop valve may be of bad design or the

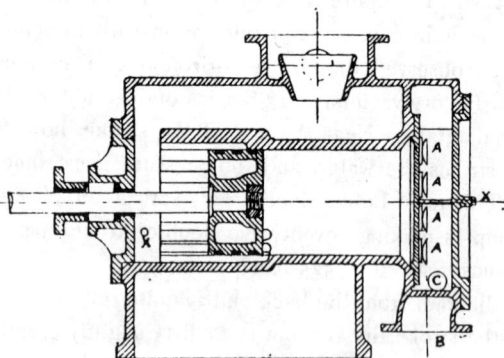

Fig. 27.

throttle valve (if any) may account for the wire drawing.
An indicator diagram taken from the steam chest of the
cylinder and another from the pipes on the boiler side of
the stop valve would show whether the stop valve were
responsible for the throttling or not.   Referring to Fig. 28,
if the diagram from the boiler side of the stop valve were
similar to the upper diagram, that is approximately a straight
line, it would show that up to the stop valve the steam pipes
were of ample capacity.   The lower figure is a typical figure
from a steam chest, which has been supplied through a

throttling stop valve. As the steam port opens, the pressure in the steam chest falls because the steam from the stop valve cannot enter in sufficient volume to keep up the pressure.

There is always a marked difference between steam diagrams and diagrams from water pumps, not only in the general form but in the very nature of the lines. In the case of steam cylinders we have smooth lines, easy and well rounded corners and smooth curves, but in pumps where the fluid is incompressible the lines are abrupt, corners are sharp

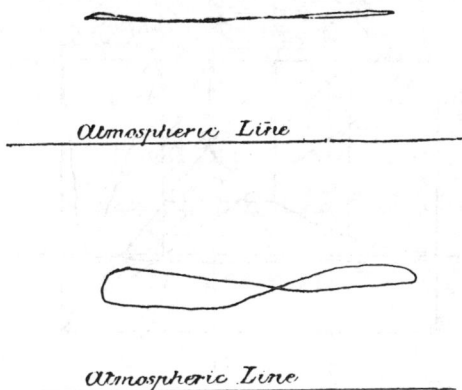

Atmospheric Line

Atmospheric Line

Fig. 28.

and the general appearance of the diagram jerky and harsh. This shows the severity of the shocks to which pumps are subject, and accounts for the great strength which is rightly employed in the construction of these machines. In connection with pumps it is interesting to observe the effect of an air vessel on the delivery pipe. Without an air vessel the diagram from a high-pressure force pump is almost a rectangle with several peaks near the corner at which delivery takes place. The effect of the air vessel is to eliminate the

peaks in a great measure and generally to smooth
the diagram. This means that the shocks are proportionally
reduced.

Diagrams from air compressors are often very instructive
and useful. It is always interesting to calculate the power
developed in the air cylinder and compare it with the work
the steam cylinders are doing at the same instant. It furnishes
a rough estimate of what the efficiency of the compressor is.

In themselves the air diagrams show if the valves are
prompt in action, whether much leakage is taking place, and

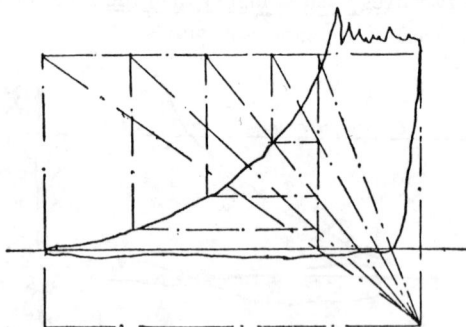

Fig. 29.

whether the valves are dancing on their faces. The accompanying air cylinder diagram (Fig. 29) is intended to illustrate
several defects to which compressors are liable. Starting at
the left-hand corner where the piston begins to compress,
the fact of the compression line coinciding with the isothermal
line shows the valves to be leaky. Isothermal compression
supposes that the fluid suffers no change in temperature
during the operation. It is an ideal condition never realised
in practice but always striven for, as witness the elaborate
cooling apparatus on some compressors. What really hap-
pens in a compressor with tight valves is a compression line

lying outside the isothermal line, but inside what is termed an adiabatic curve, which represents the behaviour of a gas under compression without it giving off or receiving any heat whatever, except that which arises solely from compression. This again is an impracticable state of things, and consequently the actual curve, as said, lies inside the adiabatic line. The nearer the compression curve approaches the isothermal the more efficient will be the compressor, but when they coincide, it is too much of a good thing and leakages are to be looked for. The pressure at the point at which the delivery valves open is higher than the receiver pressure, and in the case of a compressor with automatic valves shows

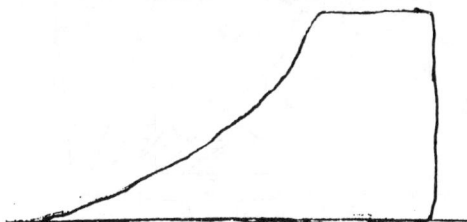

Fig. 30.

the valve action sluggish. The delivery line is jerky, and indicates dancing or flapping of the delivery valves. As the piston returns on the backward stroke the pressure in the cylinder does not fall immediately, as the curved line of the diagram shows. The inference is that either the delivery valves are leaky or the clearance is excessive. Finally, the intake of air is not satisfactory because the suction line falls below the atmospheric line. This denotes that the area through the inlet valves is too small and prevents a full charge entering the cylinder. Flapping or dancing of the suction or inlet valves would be shown by a jerky suction line on the figure. For the purpose of comparison with this very defective card a good air cylinder diagram is shown in Fig. 30.

This and the preceding chapter on indicating and indicator diagrams are intended to be suggestive rather than exhaustive; otherwise, this part of the subject would be inordinately long. The chief points have been touched upon, however, and enough has been said to enable the engineer to interpret most diagrams. When well understood an indicator diagram explains the working of steam and other engines, pumps, and air compressors more fully and more clearly than whole pages of description and illustration.

CHAPTER VII.

VALVE SETTING.

HE important subject of valve setting follows naturally after indicating, and the setting of a simple slide valve is the first operation to deal with. Sometimes the position of the eccentric is pre-determined and fixed by a key on the crank shaft, in other cases the eccentric is merely secured with a set screw, and can be readily moved to any angle relative to the crank. Taking the case in which the eccentric is keyed to the shaft, and cannot be moved, the manner of setting the valve is as follows:—Place the crank on either front or back dead centre and adjust the nuts on the valve spindle so that the valve opens the port very slightly, say, about one-sixteenth of an inch. Then bar the engine round to the opposite centre and note the amount the port is open. If correctly set it should be equal on both sides. If unequal move the nuts on the spindle half the difference between the openings so as to equalise. When there is no key on the shaft and no mark by which the eccentric may be set the operation is more tedious. Place the valve central on the face of the ports and measure the amount the valve overlaps the ports (the valve will cover up all the ports so that lines corresponding with them must be drawn outside the surface covered with the valve). This amount of overlapping is called the outside lap. Now place the crank on the dead centre and connect up the valve gear complete but do not secure the eccentric on the shaft. Put the valve gear in its central position by moving the eccentric. Then, the crank being still on dead centre, move the valve gear until the valve has moved a distance equal to the amount of overlapping as previously found : move the valve a little further so as to give the lead, say one-sixteenth, and then secure the eccentric. Then bar round and tug the lead at the other end when the crank is on the opposite centre. If equal, the valve is correctly set, any inequality is corrected by the nuts on the valve spindle. These instructions are on the supposition that the valve is properly designed and has the right travel. When the valve action is not satisfactory and it is proposed to make some alteration, it requires some thought to decide what is the best thing to do, and unless care is used the action may be worse after alteration than before.

The following list of alterations and their effects on the action of the valve will perhaps be useful :—(1) Cutting outside lap off the valve gives more port opening and later cut-off, but increases the lead seriously unless the position of the eccentric is altered. (2) Altering the angle of the eccentric and not altering the outside lap increases or diminishes the lead seriously. (3) Cutting the inner or exhaust edges of the valve gives an earlier release but less compression. (4) Putting strips on exhaust edges gives a later release and more compression. (5) Putting strips on steaming edges of valve eliminates lead unless angle of eccentric is altered; when angle is altered to suit, cut-off is earlier, exhaust is earlier, and there is more compression.

At first these statements are not very obvious. In any case they can be most clearly shown by means of a small model, in which provision for making the above-mentioned alterations is made. By this means they will become more fixed in mind.*

---

* For graphic methods of tracing the action of valve gears of many kinds readers are referred to the writer's work on " Valves and Valve Gearing."

In setting a link motion place the crank on one dead centre, connect up the eccentrics and link motion, and link up in mid-position, adjust the valve so that it is open to steam a small amount for lead, say one-sixteenth; then tighten up nuts, put crank on opposite centre and notice the lead; if unequal, equalise. Then bar the engine round with link motion in mid-position, and if correctly set the valve will open the port to steam at each end by the amount of lead. Then put link motion in full forward gear and try round, noticing particularly the variation that has taken place in the lead. Try over in full backward gear, also noticing the lead variation. Should there be no setting marks or keyways, or eccentrics, the link motion should be put in full forward or full backward gear, and the valve set in a similar manner to an ordinary slide valve gear; afterwards try over in mid and opposite gear, and correct as much of the inequalities as possible by nuts on the valve spindle. In setting the gear it is a good plan to arrange the eccentrics so that when the reversing lever is moved forward the engine runs forward and *vice versâ*. Most link motions alter the lead as the gear is moved from mid to extreme position, hence the importance of watching this point.

The Cornish valve gear of a winding engine, although somewhat complicated, is an easy one to set because of the large degree of adjustment permitted by the levers and rods. This class of gear can also be set without taking off the valve covers, etc. Cornish or double-beat valves can, of course, have no lap, but the clearance between the lifters and the spindle is the equivalent, and produces the same effect as lap on a slide valve. The opening or lift of the valves can be shown by marks on the spindle. When each valve is down mark a line on the valve spindle opposite the top of the gland from which the lift can be measured at any desired point.

Assuming first that the position of the eccentrics is fixed by the keyways, couple up the gear and put the crank on dead centre and the link motion in mid-position. Then adjust the lifters so that the steam valve at the proper end of the steam cylinder is just open for lead. Place the crank on the opposite centre and adjust lifters to give lead. Try over in full forward and full backward gear and correct any inequality as far as possible, leaving slightly more lead at the back end of the cylinder to compensate for the extra inertia on the back stroke. This completes the setting of the steam valves.

To set the exhaust valves, decide upon the point at which the exhaust valves must open. As a general rule the exhaust valves of a winding engine should close at about 95 per cent. from the beginning of each stroke, but other engines should have an earlier exhaust closure. Place the crank in the position in which it is desired the exhaust shall open, and with the link motion in full forward or backward gear, adjust the lifter so that it is just in contact with the stirrup of the spindle, put the crank in the exhaust opening position on the other stroke and set the lifter just striking the stirrup. Then bar the engine round and note where compression takes place, that is, the place at which the exhaust valves close, and which is indicated by the exhaust lifters leaving the striking part of the stirrup. Should this be considered unsatisfactory adjustment is of course possible, but it must be remembered that any alteration in the exhaust lifters or stirrups upsets the setting for the opening of the valve, just as in a slide valve cutting the exhaust edges alters both exhaust and compression. These events in fact are in sympathy with each other and cannot be altered independently. Having thus completed the setting, it is advisable to try over the complete gear for middle, forward, backward, and intermediate positions of the reversing lever. When no keys or marks are on shaft or eccentrics, the method of setting will be inferred from previous instructions, remembering that when steam lifters are in the central position clearance between them and the stirrups has ·the same effect as outside lap on a slide valve, and similarly the clearance between the lifters and stirrups of the exhaust valves is the equivalent of negative inside lap of a slide valve, whilst lift of the exhaust valves when the lifters are central is similar in effect to positive inside lap on a slide valve. These methods of setting refer to all kinds of horizontal engines except air compressors and direct-acting pumps, which, as mentioned in a previous chapter, should have no lead whatever; indeed it is better that when the crank is on dead centre the steam port should be one thirty-second of an inch closed. Compression, too, may be less than in ordinary cases.

Vertical engines require more lead and compression on the bottom side because of the weight of the parts being always downwards. Another consideration is that wear in the valve motion is in a downward direction chiefly, so that were the valve set with equal leads at first the top lead would increase with wear and the bottom lead decrease. Taking both these points into consideration, it will be good setting to have the bottom lead twice the top lead on starting, and from time to time the valve spindle nuts should be slightly adjusted to maintain this state of things. In horizontal engines wear in the valve gear brasses may disturb the valve setting. It depends on the design of the eccentric rod ends in which direction the valve is disturbed. With a strap end the valve will move towards the front of the steam chest giving more lead port opening and compression at the back of the cylinder as adjustment is made. This will be obvious when it is considered that in an end of this description the effect of drawing up the cotter is to close the strap on the outside brass and draw it towards the centre of the rod.   The other brass is solid on the end of the connecting rod at all times, and therefore cannot move outwards when adjustment is made. With an end having the cotter or wedge behind the inner brass the effect of adjustment is to force out the inner brass, the other at all times being solid against the end of the rod, hence the effect of adjustment to take up wear in the brasses is to lengthen the rod.   These remarks will no doubt suggest reasons why indicator diagrams from the same engine vary from time to time and will show the advantage of periodical indicating.

Although valve gear adjustment may account for much of this difference in the diagrams, wearing of the valve edges is the cause of some alteration.   When engines are new the valve and port edges are sharp and straight, but after a time the rush of steam wears off the sharp edges and corners, just as the cataract will wear out and smooth the solid rock on which it leaps.   When valves are tight there is little wear on the face and that is of a uniform character, but once allow a slight leakage to continue for but a short time and the wear becomes rapid and curious, and local in character.   The thin end of the wedge has got in and the rush of steam between the faces wears away the surface of the metal in grooves and channels.   This shows that small leakages should be attended to as soon as discovered to prevent their development into serious proportions.

The main valve of a Meyer expansion gear is set as an ordinary slide valve.   In order to set the expansion valves, the position of the eccentrics being known, put the crank in the position at which the normal cut-off is desired to take place.   Then adjust the nut of one expansion valve so that the port in the main valve is just closed.   Bar round until the crank is at cut-off position on the other stroke and adjust the other cut-off valve edge and edge with the port in the main valve. Then mark position of the index finger on the dial or plate stamping the degree of expansion against it. Then put the crank in the position at which the earliest cut-off shall take place, adjust the valves by means of the regulator so that they are edge and edge with the ports in the main valve and make a light scratch on the index opposite the finger. Place the crank on earliest cut-off position for the other stroke, and adjust the other cut-off valve so as to be edge and edge with the port in the main valve and make a light scratch on the index opposite the finger. These two scratches should be very close together, and midway between them make a proper mark for the mean earliest cut-off, and stamp the percentage or degree of expansion there-against. The difference between these two marks is due to connecting and eccentric rod influence and is unavoidable. If equalised now, the setting for the nominal cut-off will be disturbed. Repeat this process for latest cut-off and for as many intermediate points as may be deemed advisable. Then try over throughout the stroke, watching particularly at the earliest cut-off that the expansion valves never open the port in the main valve by their back edge, which would be most objectionable and wasteful.

To those who delight in an indicator diagram with a sharp cut-off it may be interesting to alter the angle of the cut-off eccentric with a view to giving their favourite feature. Sharp cut-offs are obtained when the valves cut off by moving in opposite directions at their maximum speed and not by one overtaking the other. It may therefore effect an improvement in some valve gears to make the cut-off eccentrics have a greater lead over the main eccentrics, though this should not be permanently altered until the gear has been thoroughly tried round to see that the range of cut-off has not been reduced, or that the cut-off valves do not re-open the steam port on the back edges, both of which events are liable to result from an alteration in the angle of the cut-off eccentric. Reversing engines having expansion gear should always be set with the expansion eccentric opposite the crank, so that backward and forward running will be identical, or practically so.

The valve gear of duplex pumps differs from most gears in that the valve gear of one side is worked from the opposite cylinder. There are two ports at each end of the cylinder whereof the outer are for steam outlet and the inner for exhaust only. The valve has no lap inside or outside, consequently when in its central position it just covers both ports. Another peculiarity of the gear is the lost motion in the valve spindle, whereby the travel of the valve is less than that of the spindle by the amount of lost motion. To set the valves, place one piston in the middle of its stroke, disconnect the link from valve rod on opposite side. Then set the valve in its central position, place valve nut evenly between the jaws or the back of the valve, screw valve rod in or out until the eye on valve rod end comes into line with eye of valve rod link, then re-connect. Repeat the operation on the other side and the valves will be properly set.

Sometimes instead of having two jaws on the valve and a nut between, the nuts are on each side of the valve and the amount of lost motion is varied by screwing the nuts on the spindle. To set the valves of a cylinder having this arrangement, place one piston in the middle of its stroke and the opposite slide valve in its central position. Adjust the nuts so as to allow about three-sixteenths of an inch lost motion on each side of the valve and the valve is set. Do not disconnect the valve motion. Repeat the operation on the other side. Should the pump give short strokes the lost motion is not sufficient; on the other hand, too much lost motion may cause the piston to strike the cylinder ends.

Sometimes the lost motion is adjustable outside the steam chest, which is an advantage in the case of large pumps. In the larger sizes cushion valves are also provided on the exhaust ports, which give additional regulation of the stroke of the pistons. Should the pistons work violently, with a tendency to strike the covers, screw down the cushion valves. On the other hand, short strokes may be remedied by easing the cushion valves.

The difficulty in setting piston valves is that the valve when in its central position completely hides the steam ports, and any lines that may be drawn from them. The usual method of setting is to make a templet or staff of the steam ports, and set the piston valves to that before putting them in the valve box; then placing the crank on dead centre the valve can be placed in position and coupled up to give lead, the crank afterwards being put on the opposite centre and the lead noted, and if necessary equalised in a similar manner to that already described. It should be remembered that any alteration to the lead produces a change in the exhaust opening and closing periods of both piston and slide valves, whether worked directly or by link motion.

The setting of Corliss gear comprises two distinct operations, the setting of the valves and rods, and the adjustment of the trip levers and the governor gear. Corliss gears are of two kinds, one in which one eccentric operates both steam and exhaust valves, and the other in which separate eccentrics are used for steam and exhaust.

Double eccentric gears may be sub-divided into those having wrist plates and those having none. The single eccentric gears have wrist plates almost invariably.

To set a single eccentric gear, place the wrist plate in its central position and couple up the exhaust rods, adjusting them so that the valves are edge and edge with the ports. Place the eccentric in middle throw and couple up to the wrist plate, but do not as yet fix the eccentric. Now put the crank on dead centre and set the eccentric about 105° in front of the crank, that is 15° past the vertical.* Adjust the steam wrist-plate rod so that lead is given at the right end of the cylinder. Fix the eccentric and then place the crank on the other centre and set the steam valve to give lead as before by adjusting the steam rod to the proper length. Afterwards bar round through a complete revolution or more, carefully watching that the valves do not uncover the steam ports by the back or idle edges. The periods of release and compression may be altered by adjusting the rods.

To adjust the trip levers, prop the governor in its top position and put the crank on dead centre. Then adjust the length of the trip rod so that the catches have just been released by the trip lever. Place the crank on the other centre and repeat the operation. Note that the trip rods are connected to the proper levers so that as the governor rises the cut-off is earlier. To adjust the length of the dashpot levers release the catches and bar the engine round until the eccentric is at its extreme position and ready to engage the catches. Now adjust the length of the dashpot rod so that the eccentric must move the catch about one-eighth or three-sixteenths of an inch before it engages with the catch on the dashpot lever. Then place the eccentric in its other extreme position and adjust the other dashpot rod in the

* This applies to those engines which have the steam eccentric rods coupled to the same side of the wrist plates as the valve lever pins, and with the wrist plate rods in tension when opening the valve. When the pins are at opposite sides, or when the rods are in compression, the eccentric position is altered accordingly.

same way. When setting the dashpot rods see that the dashpot piston is quite home against the end of the dashpot chamber. The air cushion should be adjusted so that the piston dashes inwards or outwards with as little throttling as is possible to prevent knocking. Sluggish dashpots are caused by the spring being too weak or the cushioning excessive, and in the case of vacuum dashpots by a leaky piston, excessive cushioning, or by the dashpots being too small for their work.

To set the valves of a double eccentric gear with wrist plates, place the crank on dead centre and the plate on dead centre, and the steam wrist plate in its central position. Set the eccentric in mid position and connect up the rods to the wrist plate, but do not fasten the eccentric on the shaft. Couple the wrist plate to the valve lever by the wrist-plate rods and adjust the length to the marks on the rod, now bring the eccentric on the same dead centre as the crank, then move it forward until lead is given and secure the eccentrics. The position of the eccentric will be somewhere about 35° leading the crank,* and should there be no setting marks on the wrist-plate rods this will be a good trial angle. The method of setting the exhaust valves will be inferred from previous directions; also the setting of gears without wrist plates will be understood.

In setting the governor and trip gear of double eccentric gears it is very important to make sure that the trip takes place when the governor is at its lowest position, otherwise there is danger of the live steam escaping directly down the exhaust pipe. The governors of some engines are too sensitive, and are subject to what is known as hunting. A very slight increase in the speed of the engine makes a hunting

---

* See note referring to single eccentric gears with reference to the position of the eccentric.

governor fly up to the top position, thereby closing the throttle valve or altering the point of cut-off to a greater extent than the load requires. The engine then falls below its proper speed, and the governor at once falls to its bottom position. This action is mostly present in crossed-arm governors having no dashpot; such governors ought to be supplied with one. Those governors hunting in spite of dashpots probably require adjustment of the latter. A little extra throttling of the by-pass will often eliminate the jerky action.

The opposite of a hunting governor is a sluggish one, that is, one that does not respond quickly to any change of speed. If fitted with dashpots ease the by-pass, and should that be ineffectual thoroughly clean out the dashpot and put a little soapy water instead of oil in the cylinder; this is both better and cheaper than oil or glycerine, as these latter are apt to get thick and gummy in course of time. Should the action still be unsatisfactory, thoroughly clean all the governor pins and levers, carefully removing all clogged oil from all the working parts influenced by the governor.

CHAPTER VIII.

CHOICE OF AN ENGINE.

WHEN it is desired to put down new engines or re-model old ones there is always an interesting question as to the type of engine, and a short discussion of the considerations which influence the choice may not be unacceptable to those who may some day be called upon to give their advice in similar cases.

The first question to be settled is the kind of engine that will best perform its work, and the next point is the cheapest engine that will do it, not necessarily the cheapest in first cost, but the one that will be cheapest taking into account coal consumption, attendance and up-keep, as well as initial cost.

Generally the space at disposal determines whether a horizontal or vertical engine is the better to adopt. Horizontal engines are most favoured by attendants because they are more accessible, easier to clean, and less complicated; and they also require less depth of foundation. On the other hand they occupy more ground space and the friction is greater. In verticals, the friction due to the weight of the piston and rods is thrown upon the main journals instead of on the bottom of the cylinder and stuffing boxes, and the friction of an axle is much less than the friction of a sliding-piece carrying the same weight. As for initial cost there is very little difference, the vertical engine being if anything a little more costly. The boiler pressure available will determine whether the engines are to be simple, compound, or triple expansion. Nothing is to be gained by compound engines working with a boiler pressure less than 60 lbs., and triple-expansion engines are not advisable with pressures under 135 lbs.

When an engine has to be put down to perform a certain work which in course of time may be very much increased, it is a difficult question to decide what type is the best to adopt. If the engine is to work economically at first it must necessarily become overloaded and extravagant later when the increased load is put upon it, whereas if the engine is designed for its ultimate duty it will be wasteful to start with. An excellent illustration of this matter is furnished by the fan engines at the Wigan Coal and Iron Co.'s Broomfield Pit, Standish, near Wigan, which have been previously referred to. These engines are non-condensing twin-compounds, with cylinders 26 and 46 inches diameter, 4 feet 6 inches stroke, and were set to work about 1892. At first the duty was too small, and the steam pressure in the low-pressure cylinder at the time the exhaust opened was below atmospheric pressure. As the workings in the mine extended the duty of the engine increased and better results were obtained, though the best performance would not be realised until the contemplated increase of boiler pressure was instituted. As originally designed the low-pressure cylinder had a double-ported valve giving an unvarying degree of expansion, but this was soon found to be a mistake, and expansion valves of the " Meyer " type were fitted. All engines that are called upon to perform varying duties should have variable expansion gear on the low-pressure cylinder or they will at times be working at great disadvantage. In the Broomfield engines before the expansion gear was fitted the 46 inches low-pressure developed at one period only about

12 horse-power at thirty-eight revolutions and was for a considerable part of the stroke a drag upon the high-pressure cylinder, a loud knock in the crank pin brasses indicating the instant at which the force urging the low-pressure piston became a negative quantity.

In cases where the boiler pressure will at some future time be increased but where the engine must at first develop the same power as when the increased pressure is supplied, a good plan is to put down a pair of high-pressure engines to work with the low-pressure, making provision in the design of the crank shaft, the valve gear, bedplate and foundations, for the displacement of one of the cylinders by a new low-pressure of dimensions suitable for the increased pressure. This requires a little care in the drawing office, but when thoroughly well planned the alteration is not costly and the engine can be converted in a few days.

Where the boiler pressure is permanent but the load will increase, the modern plan is to put down a single or tandem-compound triple-expansion engine as the boiler pressure calls for, of sufficient power to perform the initial duty, leaving a preparation on the crank shaft for a duplicate crank and room in the engine-house for duplicate engine to be installed when the work becomes too heavy for the first engine. When properly arranged the second engine can be erected completely and coupled to the existing engine at a week-end.

For such a duty as electric lighting the modern practice is to install a number of small independent units, which may be added to from time to time as circumstances demand. The advantage of this plan is that each unit is working at its most economical load besides the chances of complete failure of the system being very remote.

The considerations which determine whether an engine shall have one, two, or more cranks may be the space at disposal and the nature of the work to be performed. In some cases it is an imperative condition that the engine shall be able to start at any position. Winding and hauling engines are instances. Here two or more cranks are indispensable; at other times questions of duplication may decide the matter. Where neither ability to start from all positions nor the question of duplication are concerned, nor the question of space, regularity of running is generally the deciding factor. The twin-compound engine is steadier running than the tandem-compound and the cost is not much higher, for although there are two connecting rods, two bedplates, and double sets of eccentrics and rods in the twin engine, all these parts are much lighter, as well as the crank shaft and flywheel, consequently for the driving of a textile factory or for generating electric current for lighting purposes when uniform speed is important, the two or three-crank engine is nearly always chosen, space permitting.

The pros and cons concerning fast and slow revolution engines are as follows:—High-speed engines are smaller and therefore cheaper than slow-running engines of the same power, and the percentage of speed variation is less. On the other hand they are not so durable and require more attention, nor can they be fitted with certain forms of valve gear which have proved themselves most efficient.

Whether engines shall be condensing or non-condensing is a question of the water supply, except of course in the case of intermittent running and very small engines which are mostly non-condensing.

Whether they shall have jet condensers or surface condensers depends on whether the hotwell discharge is used for boiler feeding, and if so, whether the water contains matter deleterious to the boiler plates. The advantage of the surface condenser is that except for the small amount of supplementary feed that must be made up to compensate for leakages, the water is in constant circulation through the engine and boiler, and consequently there is no accumulation of sediment and scale. With the jet condenser fresh water continually enters the boilers, and if impure, is constantly depositing scale or mud therein. In other respects the advantage is decidedly with the jet condenser. It is both simpler, cheaper, and easier to superintend. The position of jet condensers is determined by the level of water from which the injection is drawn. They will not draw their water freely with a lift of more than about 15 feet, so that sometimes the pump has to be placed low down and a long rod employed to work the air pump.

Another form of condensing apparatus is in use in situations having only a limited supply of water, but where the amount of power enveloped is of such magnitude as makes the saving effected by condensing amply compensate for the cost of the plant. The arrangement consists of a cooling tower to the top of which the condensing water is lifted; from thence it falls upon pipes containing the exhaust from the engines which is thereby condensed. When the water has reached the bottom of the tower it is warm, and before being put into circulation must be cooled. Various means of doing this are adopted, the common plan being to spray it over the surface of a shallow tank. Examples of this form of condensing apparatus are mostly seen in connection with the electric stations in the centre of large towns.

Independent condensing plants are of advantage in some situations. Where there is a number of non-condensing engines in close proximity and it is desired to make them condensing, it is found better to put down a condenser common to all, and employ a small steam engine for the sole purpose of driving the air pump. The advantages of this system will be obvious, and apart from considerations of cost, each engine is enabled to commence work with a good vacuum from the start.

In the matter of valve gears there is a wide choice at the engineer's disposal, and it may not be out of place to discuss some points of this important part of the subject. Were the slide valve an easy one to work under high pressures and with large sizes, it would not be easy to devise anything to equal it. It has this advantage—which is not possessed by some other types—that the effect of wear has not the tendency to increase leakage but rather to diminish it. This feature is absent in piston valves. An engine with slide valves will often give as good a performance twenty years after it has been built as on the day it started, whilst some vertical high-speed engines give fair results at first but gradually depreciate until they become very wasteful through wear in the valves. Cornish valves are still employed largely in winding engines, but it is not claimed that as commonly arranged they are economical. The feature which recommends them is the ease with which they may be handled on the reversing lever. Corliss valves have as good features the effective drainage of the cylinder combined with separate steam and exhaust ports and small clearances. The gear is complicated and requires a lot of attention, but its action is almost perfect, and it permits of very sensitive governing. Corliss gear is mostly confined to large slow continuous-running engines, though latterly it has been applied to winding and hauling engines, especially in foreign gold mines, where fuel is dear. Corliss valves do not depreciate with wear, but the gear is soon liable to become noisy and loose unless the best attention is given. Many Corliss gears are altogether too light for their work, and constant renewal of the pins and bearings is necessary. This has brought discredit on this class of engine in some quarters, and the class has been condemned on the performance of the particular example. Corliss engines should not be advised where the most skilled attendance is not procurable, and where the situation is dusty. For such conditions enclosed engines are the most suitable, as they require little attention and are dustproof.

CHAPTER IX.

CONCLUDING REMARKS.

WHEN it is proposed to increase the boiler pressure of an existing plant there is always considerable responsibility upon the shoulders of the engineer, who has to advise whether the present engines are good for the increased load. To decide this question requires a knowledge of the method of calculating the strength of various parts of machines, to treat of which is not the province of this little work. It will, however, be useful to indicate where the weak points are chiefly to be found, and a few of the considerations which will guide the engineer to his judgment.

With regard to the cylinders, steam chests, pipes, and bedplates, calculations will not be of much use, because in the first place it is almost impossible to say what thickness the metal is in the thinnest and weakest parts; and again, the nature of cast iron is so uncertain that figures attempting to deal with it are at best founded on many assumptions. This may seem a very off-hand way of dealing with a subject, but fortunately it is not these parts by which the strength of an engine is to be measured. That they are subject to failures is true, but in the great majority of cases failure is through shocks that are not incurred in the course of ordinary working, and the probability is that they would have fractured had they been 30 per cent. stronger.

Steam pipes, for instance, rarely burst by steam pressure alone, but generally through water-hammer, accidental blows

and strains set up by expansion. The same remarks apply
to cylinders and steam chests, whilst bedplates are most frequently broken through sinking of the foundations or water
in the cylinders.

The piston rods are the first details that should be in-
spected. The weakest spot will either be in the thread of
the piston nut, or through the cotterhole in the cross-head.
It is suggested that the engineer, before going over the
engine in detail, should prepare a table in which particulars
of the various parts can be entered and afterwards compared with one another. The arrangement of the table
admits of many variations, but the accompanying specimen
includes the particulars which should be entered.

Proposed increase of Steam Pressure from _____ lbs. to _____ lbs.

Description of Engine _____

Diameter of Cylinder _____ ins.

Stroke _____ ft. _____ ins.

Part.	Area of section under stress.	Maximum power to be sustained under increased pressure.	Stress per sq. inch.	Nature of stress.	Remarks.
Through the cotter-hole of piston-rod.	7 sq. ins.	42,000 lbs.	6,000 lbs.	Alternate tension and compression with shocks.	Cotter-holes rounded. No appearance of cracks. Mild steel rod.
Thread of piston-nut.	8 sq. ins.	42,000 lbs.	5,250 lbs.	Do. do.	Good thread and nut. Wrought-iron nut. Steel rod.

The piston rod of an engine is subjected to alternate

G

tension and compression, violent shocks, and rapid changes of load. Now, if the rod be made of mild steel with a tensile strength of 28 tons per square inch (which is the average strength of good mild steel), a factor of safety of fourteen may be allowed. That is to say that the rod must be made of such strength that it would require theoretically fourteen times the load to which it is subjected in order to break it. At first this will be thought an absurdly wide margin to allow, but when all things are taken into account it will not appear so preposterous. First there is an allowance to be made for any internal flaws in the material, then there is the violent and rapid change in the nature of the stress to be taken into account. Over and above this there must be an allowance to meet accidental shocks, such as water in the cylinder, and when the rod is in compression it must be quite rigid.

Taking the concrete case of a condensing engine with a cylinder 24 inches diameter, and working with boiler pressure of 100 lbs., what diameter of piston rod will be required? The thrust on the rod is $452 \times (100 + 15) = 51,980$ lbs. The strength of the steel being 28 tons per square inch, and allowing a factor of safety of fourteen, the area of the rod must be $\dfrac{51980 \times 14}{28 \times 2240} = 11\cdot6$ square inches. This must be the area through the weakest part of the rod, usually the cotterhole for securing the crosshead to the rod, and as this is generally about 35 per cent. less than the full area of the cross-section of the rod, therefore the area through the solid piston rod will be 15.6 sq. inches, which corresponds to a diameter of 4½ inches.

Similar calculations determine the diameter of the connecting rod, but since this part is subject to a strain due to the vibration an extra allowance must be made. This stress is not of much account in a slow-running engine, but at high

speeds it becomes important, hence the connecting rods of
many high speed engines are made of rectangular section
so as to offer more resistance to bending due to the vibration of
the rod itself.   The body of the connecting rod is not usually
weakened by a cotterhole, hence it is a common practice to
make the small end the same diameter as the piston rod,
tapering towards the crank end to a degree that pleases the
eye or the fancy of the designer.

The crank shaft requires careful designing to allow for the
various strains which come upon it; the forces acting upon it

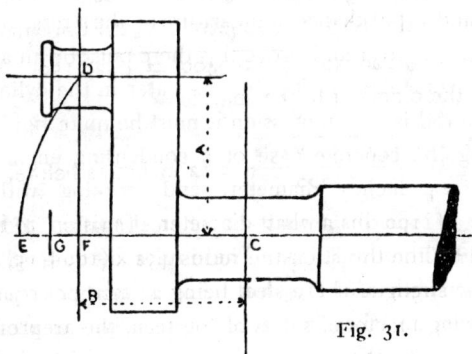

Fig. 31.

are a twisting force, and a bending force due to the overhang
of the crank and the pressure of the piston. It will be
convenient to take the same case as when dealing with the
piston rod. Let Fig. 31 represent the crank and shaft of
the engine and let the distance A = 30 inches and B = 25
inches. The shaft must resist a twisting moment equal to
30 × 51,980 = 1,559,400 inch-lbs., and a bending moment
equal to 25 × 51,980 = 1,299,500 inch-lbs., and must be
strong enough to resist the combined action of the two.   The
following is a simple graphic method of determining the
diameter of the crank shaft and one which is thoroughly

reliable.   From C as centre, and with CD as radius, describe an arc DE, draw the vertical DF, and bisect EF in G.   Then, 51,980 lbs. being the thrust on the piston, the diameter of the shafts will be

$$\sqrt[3]{\frac{51980 \times \text{distance CG in inches} \times 10.2}{10000}}*$$

Distance CG is 33 inches, hence the diameter of the shaft is 12 inches nearly.

Solving the problem by another method, the twisting moment is 1,559,400 inch-lbs., and the bending moment is 1,299,500 inch-lbs.   The equivalent twisting moment = 1,299,500 + $\sqrt{1299500^2 + 1559400^2}$ = 3,428,150.   Then by this rule the diameter of the shaft is

$$\sqrt[3]{\frac{3428150 \times 5.1}{10000}} = 12 \text{ inches as before.}$$

The factor 10,000 is a constant for steel shafts.   If the shaft be wrought iron use 8,000.   These rules are quite satisfactory for the strength of shafts, but there is another consideration which must be taken into account, namely, the pressure per square inch on the bearing due to the weight of the shaft itself and the weight of the driving pulley and fly-wheel as well as the weight of the cranks, connecting rods, eccentrics, etc.   This consideration is too frequently neglected in works on the steam engine, yet the point is a most important one. Experience has shown that satisfactory working cannot be relied upon if the pressure per square inch greatly exceed 160 lbs.   Thus, if the whole weight of a crank shaft be 40,000 lbs., and assuming the fly-wheel be central, the pro- jected surface of each bearing must be $\frac{20000}{160}$ = 125 square

---

* This construction is due to Mr. C. N. Pickworth.

inches. (By projected surface is meant the product of the diameter and the length : therefore in the present case a bearing 9 inches diameter × 14 inches long would give the necessary surface.) In some single engines the fly-wheel is placed near the outer bearing which has thus to carry more than its share of the dead load, and unless this bearing is of liberal proportions trouble will be experienced. Almost every engineer has at some time or other had trouble of this nature.

In crank pins the dead weight is inconsiderate, and strength being satisfied the bearing should be designed so that the maximum pressure upon them due to the thrust of the piston does not exceed from 600 to 1,000 lbs. Such wide limits may be thought very indefinite but they include extremes. An intermittent-running engine, such as a winding engine, would run satisfactory at 1,000 lbs. per square inch, but for a fan engine such a pressure would be altogether too high, and a bearing pressure of 600 lbs. per square inch would not be too little. These two classes of engines represent extreme cases from which suitable pressures for other types of engines may be deducted. As to the strength of the crank pin it may be calculated as an overbearing beam on cantilever of circular section, with a concentrated load equal to the maximum thrust on the piston rod acting on the central point. The motion on the crosshead pins being slight, pressures up to 1,400 lbs. per square inch may be safely allowed. However abundant the proportions of wearing parts of engines, trouble will arise, unless the bearings are in line and free from grit.

In compound and triple-expansion engines the thrust on the piston rods is, of course, measured by the difference in the pressures between the steam on each side of the piston, hence in calculating whether engines of this type will safely sustain an increased boiler pressure the probable pressures in the cylinders at each side must be most carefully considered.

Superheated steam after being discarded for some fifty years is coming into use again, and considerable economy has of late been obtained with engines working on this system. There is no special difficulty attended with its use, the valves being the parts chiefly concerned. At the high temperatures which obtain, abrasion of the faces of the valves having a sliding motion is liable to occur, hence Cornish or drop valves are frequently adopted, and are found to give satisfaction when proper gear is employed to work them. With superheated steam the engineer must use increased vigilance. All surfaces about the cylinders require to be very well clothed. Special attention must be given to the effect of expansion on the various parts, and the various fittings about the cylinders ought to be of the very best description. Any soldered joints in the lubricators and taps are inadmissible because the solder melts with the heat : with regard to the superheaters themselves, they are sometimes placed in the flue at the back of the boilers and sometimes they are separately fired. The pipes should be examined weekly and all exterior deposits removed. The relief valve should be lifted occasionally to guard against sticking. The thermometer sockets are liable to become choked up with dust and should be cleared out weekly. Frequent reading of the thermometer will give a good idea of how the superheater is working and whether its efficiency is deteriorating. Special precautions must be made that no timber work is in close proximity to the steam pipes.

INDEX.